I0463081

(Science Primers.)

PHYSICS

BY

BALFOUR STEWART, F.R.S.

Late Professor of Physics, Victoria University, The Owens College,
Manchester, Author of "Elementary Lessons in Physics."

WITH ILLUSTRATIONS.

London:

MACMILLAN AND CO.

AND NEW YORK.

1890.

PREFACE.

IN publishing the Science Primers on Physics and Chemistry, the object of the Authors has been to state the fundamental principles of their respective sciences in a manner suited to pupils of an early age. They feel that the thing to be aimed at is not so much to give information, as to endeavour to discipline the mind in a way which has not hitherto been customary, by bringing it into immediate contact with Nature herself. For this purpose a series of simple experiments has been devised, leading up to the chief truths of each science. These experiments must be performed by the teacher in regular order before the class. The power of observation in the pupils will thus be awakened and strengthened; and the amount and accuracy of the knowledge gained must be tested and increased by a thorough system of questioning.

The study of the Introductory Primer will, in most cases, naturally precede that of either of the above-named subjects; and then it will probably be found best to take Chemistry as the second and Physics as the third stage.

The whole of the apparatus needed for all the experiments (except a few marked in the text with an asterisk) will be supplied by the firms whose names may be found at the end of this Primer, for £19 3s. 8d. exclusive of packing.

If prompt payment be made, these firms are willing to supply the set for £17 exclusive of packing cases, which may be either 9s. or 21s. according to quality. Those who have already purchased the apparatus for the Chemical Primer need not purchase another Grove's battery.

TABLE OF CONTENTS.

SCIENCE PRIMERS.

PHYSICS.

INTRODUCTION.

1. **Definition of Physics.**—You have been told in the Chemistry Primer what sort of things we have around us. You have seen what the chemist does; how he weighs things and finds their quantity, and also how he finds that certain things are compound, and may be split up into two or more new things; while again other things are simple or elementary and cannot be so split up.

In fact you have been told about the various **kinds** of things we have in the world, but you have not yet learned much about the **affections or moods** of these things. You are yourself subject to change of moods; sometimes you appear with a smile on your face, and sometimes, perhaps, with a face full of frowns or tears; sometimes, again, you feel vigorous and active; sometimes dull and listless.

Now if you think a little you will see that the

ℭ

things around you are subject to moods very much like yours. To-day the face of nature looks bright and happy, and full of smiles; to-morrow the same face is dark and lowering; the rain falls, the thunder roars, and the sea is tossed with waves and very stormy. Or again: let us take an iron ball which lies upon the floor; it is cold and heavy to the touch, but let us put it into the fire, and when it comes out the same substance is there, but the state of it is very different; if you now attempt to touch it, you will be sure to burn your fingers. Or again: if, instead of putting it into the fire, we put it into a cannon and discharge the cannon, it will come out with tremendous velocity, and will knock to pieces anything it touches.

Thus you see that a cold cannon-ball is a very different thing from a hot cannon-ball; and also that a cannon-ball at rest is a very different thing from a cannon-ball in motion.

Now if we see you crying and unhappy, we ask what is the cause of this mood, and we always find there is a cause; or if we find you listless and sleepy, and wanting energy, we inquire what is the meaning of all this, and we find that it has a meaning and a cause. So likewise when we find changes in the moods or qualities of dead matter we inquire what is the cause of these changes, and we always find they have a cause. This inquiry we shall make in the following pages, and you must attend well to the answer we get. You have already been told that this mode of questioning nature is called **experiment**.

2. **Definition of Motion.**—You must in the first place get a clear idea of motion. Motion means change of place. Some of you may have heard that this solid earth on which we dwell is in truth moving very fast round the sun, but we may, in the meantime, put away this thought altogether from our minds, because although the earth is moving very fast it carries us all along with it, and everything goes on as smoothly and quietly as if the earth were at rest.

Well then, if I sit on a chair in a room I may say that I am at rest, but if I walk up and down the room I am in motion. Now in order to understand my movements, you must know something more than the mere fact that I am moving about; you must know the direction or line in which I am moving, and you must also know the rate or velocity with which I am moving. You must try clearly to understand the meaning of this word " velocity "; and to make you do so, let us suppose that I go out of doors and walk along a straight road for two or three hours, and always at the same pace. Well, I find that in one hour I have got four miles beyond my starting point, and that in two hours I have got eight miles beyond it, and I therefore say that I am walking on at the rate or with the velocity (for both words mean the same thing) of four miles an hour.

But what if the rate be not always the same. Suppose a railway train to be coming near a station, and just beginning to slacken its speed. The train is first of all moving, let us say, at the rate of forty miles an hour, but presently its velocity gets less and less, until when it arrives at the station it comes quite to a standstill. Now, how can we find its rate when this is always changing? or what do we mean when we say that the train, before it began to slacken its speed, was moving at forty miles an hour? We simply mean, that if the train had been allowed to move for a whole hour at the same rate it had before it began to slacken its speed it would have moved over forty miles. In fact, if instead of coming to rest at the station it had been an express train, and gone on, it would have been forty miles away an hour after we began to notice it.

There are different ways of expressing velocity : sometimes we speak of so many miles an hour, as we have done here, but sometimes it is better to use feet and seconds ; thus if I drop a stone down a well I should say that it fell sixteen feet during the first second after it was dropped. Sixty seconds, you all know, make a minute, and sixty minutes make an hour.

In this little book, when speaking of velocity or rate, we shall use feet and seconds more frequently than miles and hours, and speak of a body as moving at the rate of ten, or twenty, or thirty feet a second, as the case may be.

3. **Definition of Force.**—Now what is it that sets in motion anything that was previously at rest ? Or what is it that brings to rest a thing that was previously in motion ? It is force that does this. It

is force that sets a body in motion, and it is force
(only applied in an opposite direction) that brings it
again to rest. Nay, more, if it requires a strong
force to set a body in motion, it requires also a strong
force to bring it to rest. You can set a cricket-ball
in motion by the blow of your hand, and you can also
stop it by a blow, but a massive body like a railway
train needs a strong force to set it in motion, and a
strong force to stop it. That which is easy to start
is easy to stop; that which is difficult to start is
difficult to stop. You see now that force acts not
only when it sets a body in motion, but as truly when
it brings a body to rest. In fact that which changes
the state of a body is called force, whether that
state be one of rest or of motion.

EXPERIMENT I.—To prove this, take a tin pan with
some peas in the bottom of it, and hold the pan in
your right hand. Now quickly raise your right hand,
with the pan in it, until your right arm is brought to
a stop by a fixed bar of wood, which you have placed
a little above it (your other arm held stiffly will do as
well as the wood). Now what you have done is to
make the pan with the peas rise quickly up, and then
suddenly come to a dead stop. You have first, by the
force of your arm, given an upward motion to the
pan, and the pan has forced the peas to mount with
it, since clearly they could not remain behind. Then,
again, when your right arm holding the pan was
mounting quickly, you allowed it to be stopped all at
once by the bar of wood; that is to say, the bar of

wood forced your hand to stop, and your hand in its
turn forced the pan, which you held tightly, to stop
also. But this stopping force does not affect the peas
which lie loosely at the bottom of the pan, so that
they will continue to mount up after the pan has

Fig. 1.

been stopped, and many of them will fall over the
edge and be scattered about upon the floor.

EXPERIMENT 2.—Now put some more peas into the
pan, having spilt the last ones ; but instead of raising
the pan quickly upwards, lower it as quickly as you
possibly can. Here, the force of your arm makes
the pan move down very quickly, but does not
affect the peas which lie loosely on the bottom of
the pan; the result will be, that the peas will not
follow the quick motion of the pan, but will lag
behind until at last they are all scattered about upon
the floor.

Let us now pause for a moment, and see what we
really learn from these two experiments. We learn

from the first, that after we have once set the peas in motion upwards, since the stopping force of the bar of wood does not affect them, they continue to move upwards after the pan has been stopped. It requires force to stop their upward motion, and this force we could not apply by means of the bar of wood, so that they continue to mount upwards until the force of the earth at last brings them downwards to the floor. You see, therefore, that it needs force to stop a moving body.

Again, in the second experiment, we communicate a downward motion to the pan, but the force of our arm which does so, does not affect the peas which lie loosely on the bottom of the pan. They, therefore, keep their state of rest, and lag behind the pan until at last the force of the earth brings them downwards to the floor. You see, therefore, that it needs force to start a body at rest.

Force, therefore, may do two things; it may either stop a body in motion, or it may set in motion a body at rest. But very often we find that a force, although present, does not appear to act. Now, why is this? We reply, because it is prevented from doing so by another equal and opposite force. Thus, I hold a heavy weight in my hand; if I open my fingers, the force of the earth which acts upon the weight will bring it very soon to the floor; but as long as I keep my fingers shut I prevent this force from acting. Or, imagine the same weight to lie on the table; if there were no table, it would fall to the floor; but the force of the earth which gives it a tendency to fall, is prevented from acting or is resisted by the table. The weight presses against the table, but the table withstands this pressure. So that you have here two forces resisting or withstanding each other, the one being the weight, and the other the resisting force of the table.

From all this we learn that force is that which changes the state of rest or of motion of a body, but that very often force is resisted or prevented by an equal and opposite force, so that it is not able to do anything or to produce any effect.

THE CHIEF FORCES OF NATURE.

4. Definition of Gravity.—I have thus told you what is the meaning of the word force, and now let us look about us in order to see what are the chief forces with which we have to do, and to see also what part each plays, and what is its use. The most prominent force is the attraction of the earth. If we let go a heavy thing out of our hands, we know where to look for it; we know that it will not mount towards the sky, nor will it move off sideways in some direction, but it will fall to the ground or earth. It falls down, we say, and the very words up and down depend upon the earth's force; so that if the earth had no force, we should not use such words at all. The word "up" denotes a difficult motion against the earth's force; the word "down" an

easy motion, by help of the earth's force. It is difficult to walk up a hill, but it is very easy to walk down.

Now when we say that the earth attracts things, you must not think that all, or nearly all of the things which we see are moving towards the earth. You and I are not so falling, nor should we wish to be in such a very dangerous condition. Why are we not falling? Because we stand upon the floor; but if there were no floor, we should fall through to the ground, and the floor must be strong enough to support our weight, otherwise it would give way and we should fall. Sometimes a wooden floor or platform has been so filled with people that it has given way, and they have fallen to the ground, and many of the people have been killed or very much hurt.

Thus you see that the earth attracts everything, but yet most of the things which we see around us are not moving towards the earth, because they are supported by something else that is able to resist their weight. In fact, this property of things called weight is really caused by the attraction of the earth.

This force which the earth exerts is called **gravity**.

5. **Definition of Cohesion.**—But there are other forces besides that which the earth exerts. If we take a piece of string or of wire, and try to break it into two parts, it exerts a force to prevent our doing so, and it is only when the force we exert is greater than the force with which it resists us that we succeed in breaking it. In fact the different parts or particles of the string or of the wire are held together by a force

which resists any attempt to pull them asunder. And
so are the various parts or particles of all solid bodies,
such as wood, stone, metals, and so on. It is often
very difficult to break a substance to pieces, or bend
it, or powder it, or alter its shape or size in any way.
Now that force which binds together the various
particles of a body is called cohesion.

You see now the difference between gravity and
cohesion ; gravity is that force which the earth exerts
to pull bodies to itself, and which acts at a great
distance ; so that for instance, the moon, which is
240 thousand miles away, is attracted by the earth.
Cohesion again is that force which the neighbouring
particles of a body exert to keep each other together,
but this force does not act except when the particles
are very near each other ; for if once a thing is
broken or ground to powder, its particles cannot
come easily together again.

6. Definition of Chemical Attraction.—Besides these two forces there is the force of chemical
attraction or affinity. You are told in the
Chemistry Primer (Art. 4) that the two things coal
and oxygen gas unite chemically together, and that
carbonic acid gas is the result of their union. The
coal and the oxygen gas are pulled together by a
force which they exert on each other as truly as
a stone is pulled towards the earth. In virtue of
this force they rush together and unite, and the
result is something quite different from either. This,
then, is the force which we call chemical attraction,

and which has this peculiarity, that it can only be exerted by different bodies; for in chemistry it is only bodies of different kinds that rush together and unite after this fashion.

7. **Use of these Forces.**—Having now told you something about the chief forces of nature, let us try to find what part they play, and why they are there at all; and I think we shall soon see that we should get on very badly without them. Let us begin by supposing that there was no such thing as gravity, and that the earth did not attract things to it. Now sometimes when we climb a steep hill we are tempted to think how pleasant it would be if we could go up as easily as we go down. How we wish there was no gravity! But it would be a terrible misfortune if one of those spirits we read of were at once to grant us our request. There being no gravity there would of course be no weight, and we should then get up a hill easily enough, but if we jumped into the air we should remain there; and possibly we might be able to leave this world altogether. The furniture of our houses would be found some on the floor, some on the roof, some floating about, and we ourselves could walk on the roof as easily as on the floor. The moon meanwhile, not being bound to the earth, would leave us for ever; and in like manner the earth, being no longer bound to the sun, would leave it far behind and wander off among the stars.

So much for gravity. Let us now see what would happen if there were no cohesion. If this force were absent, the particles of solid bodies would not adhere to one another, but they would all fall to pieces or rather to powder. The wood of our tables and chairs would fall to powder, and we should have no furniture ; and the bricks of our houses would do the same, so that we should have no houses. We should do the same ourselves, and in fine, all things would resolve themselves into a huge mass of dust.

Finally, let us think what would happen if there were no such thing as chemical attraction. In the first place the fire would cease to burn because the carbon of the coal would no longer care to unite with the oxygen of the air.

In the next place no two simple or elementary substances would unite together to form a compound substance, but we should have nothing but about sixty simple substances consisting of a great number of metals and a small number of gases. There would be no variety in such a world, and indeed there would be no living in it, for our own bodies are compound ; and if chemical affinity were destroyed part of them would go up into the air and mix with it, while another part, consisting of a quantity of carbon, a little phosphorus, and some one or two metals, would fall to the ground, and thus we should come to an end.

HOW GRAVITY ACTS.

8. **Centre of Gravity.** EXPERIMENT 3.—Let us now endeavour to find out what sort of a force gravity is, and for this purpose let us take this irregular flat sheet of iron and hang it up by a thread. You see it hangs in a particular way, and you also see that the line already drawn on the sheet is in the same direction as the line of the thread, and

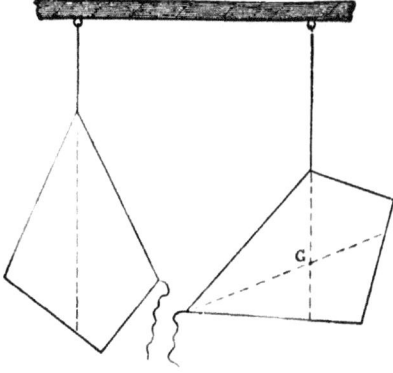

Fig. 2.

forms a vertical prolongation of it. Next let us hang the sheet freely from some other point ; here again you have another line in prolongation of the thread, and you further see that these two lines cut each other in a point marked G.

Now let us hang up the sheet by some third point

in its rim.   As before, you have a line in prolongation
of the thread.   Now you will easily see that these
three lines all cut one another at the same point G ;
in fact, if you suspend the sheet from any point
freely by a thread, and draw a line in prolongation of
the thread, all such lines will cut one another in the
same point G, so that this point will always be directly
under the point from which the sheet is hung, and if
you push the sheet to one side it will return again to
its old position.   Now what is this peculiar point G ?
To find out let me attach a string to the point G, and
hang the sheet by the string ; you see that the sheet
will balance round G in all directions just as well as
if its whole weight were condensed into the point G.
Now G is what we call the **centre of gravity** of the

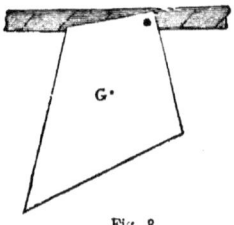

Fig. 3.

sheet ; and if I hang up the sheet freely from its
corner as in Fig. 2, it will put itself in such a position
that its centre of gravity G shall be as low as possible.
Or if instead of hanging the sheet by a string I
suspend it loosely upon a peg, it will still try to place
the point G as low as it possibly can, and it will not

hang as in Fig. 3, but the point G will place itself vertically under the peg. This centre of gravity is in fact the point at which, practically speaking, we may imagine the whole weight of the body to be concentrated and to act. Now owing to the earth's attraction a body naturally places itself as low as possible, that is to say, it places its centre of gravity as low as possible.

9. **The Balance.**—Every substance has a point G of this kind, which we call its centre of gravity. The balance which you see on page 40 has, like everything else, its point G—its centre of gravity. And it will endeavour, just like the sheet of iron, to place this point as low down as it possibly can.

Now when there are equal weights in both scale-pans, this point G at which, practically speaking, we may imagine the whole weight of the balance to be concentrated is somewhere in the pointer directly under the point upon which the balance is swung ; and hence, if by pushing I try to tilt the balance to a side, when freed it will ultimately return to its old position. In fact, when the weights in each pan are equal, it will always keep this position, with the pointer pointing exactly in the middle ; so that if I am weighing a substance, and place this substance in the one scale-pan, and the weights in the other, and if the pointer points exactly in the middle, I am then quite sure that the weights in the one scale-pan are exactly equal to the weight of the substance in the other   But if the weights are not heavy enough the preponderating weight of the substance will tilt the beam and pointer of the balance over in one direction ; while if the weights are too heavy their preponderating influence will tilt the beam and the pointer over in the other direction.

EXPERIMENT 4.—Suppose that I put this piece of metal into one of the scale-pans, and put weights equal to 150 grains into the other, the scale-pan with the metal in it sinks down, thereby showing that the metal is heavier than the weights. Next let me put weights equal to 250 grains into the other scale-pan. Now again it is these 250 grains that are too heavy, and you see that the scale-pan containing them sinks down, whereas before it was the other that sank. Thus the weight of the metal is somewhere between 150 and 250 grains. Let us therefore try a 200-grain weight, and you see that now the pointer points exactly in the middle, and the beam of the balance is exactly horizontal, showing that the weight of the metal is exactly 200 grains.

THE THREE STATES OF MATTER.

10. You have seen that we cannot do without the various forces of nature, and that if one piece of matter were not drawn or attracted to another piece, there would be no such thing as a world at all. You have seen, too, that if there were no cohesion, there would be nothing but powder. I may now proceed to tell you that if everything possessed cohesion to a

STATES.] PHYSICS.

great extent, we should be nearly as badly off, for we should in such a case have neither liquids nor gases, neither water nor air.

The particles of a bar of iron or steel possess very great cohesion, and it is very difficult to force them apart. But water and mercury have hardly any cohesion whatever, and the very slightest touch will scatter in all directions a quantity of water or of mercury. Yet these two liquids have still a little cohesion left, as you may see by the following experiments.

EXPERIMENT 5.—Take a very small quantity of mercury from the bottle containing it, and put it on a flat glass surface. By pressing it you may split it up into small globules. Now these globules are a proof that the particles of mercury cling together. For, put another plate of glass above them, and you may by this means squeeze them flat ; but if you take away the glass, the mercury will resume its previous globular shape.

EXPERIMENT 6.—Sprinkle a few drops of water on an oily or greasy surface, and these will be found to have a rounded form, not unlike drops of mercury, showing that the particles cling to one another.

On the other hand, the particles of gases, such as the air we breathe, have no tendency to keep together, but rather the reverse. Indeed they will separate from one another unless there is some force which keeps them from doing so.

So that, you see, we have three very different states

cf matter, the solid, the liquid, and the gaseous ;
and each of these states has certain properties which
serve to distinguish it.

11. **Definition of Solids.**—A solid body, such as
a piece of iron or wood, resists any attempt to alter
either its shape or its size, always keeping the same
size or volume and the same shape, unless it be
violently destroyed.

12. **Definition of Liquids.**—A liquid like water,
when kept in a bottle or other vessel, always spreads
itself out, so as to make its surface level, but yet it
will always keep its proper size or volume. You
cannot by any means force a pint of water into a
half-pint measure ; it will insist upon having its full
volume, but it is not particular as to shape.

13. **Definition of Gases.**—A gas again has no
surface ; for if you put a quantity of any gas into a
perfectly empty vessel, the gas will fill the whole
vessel. Nor does a gas insist so violently as a liquid
upon occupying a certain space, for by means of a
proper amount of force I can compress the gas which
now fills a pint bottle into half a pint, or even into
less space, if I use sufficient force. In fact, a gas will
be persuaded to go into less space, but a liquid will
not be persuaded.

PROPERTIES OF SOLIDS.

14. The peculiar distinction of a solid is that it insists upon keeping not only a certain space or size for itself, but also a certain figure or shape.

* EXPERIMENT 7.—In Fig. 4 (upper figures) you have two vessels of **different shapes,** but of the **same size.** And if you exactly fill the one with water and pour it into the other, you will find that the water exactly fills it also.

Fig. 4.

Here, again (lower figures) you see two pieces of wood that have both the **same shape** or figure, but the one is much larger than the other—their **size is different.**

You see now what is meant by space or size or volume (for the three words mean the same thing), and what by figure or shape. Now you cannot take

a solid which has the shape of the one bottle and
force it into the shape of the other, although the size
or volume of both is the same ; nor can you take a
solid of the size or volume of the first wooden block
and squeeze it into that of the second, although the
shape of both blocks is the same. A perfect solid
will keep its figure, and it will also keep its size.

Bear in mind, however, that when we say we can
not do a thing, we really mean we cannot do it without very great difficulty, and then not completely,
but only to a very small extent ; in fact, what we
really mean is best explained by making a series of
simple experiments.

EXPERIMENT 8.—Let me take a bar of iron ; I will
first of all try to break it in pieces by means of a
blow, but it won't be broken.

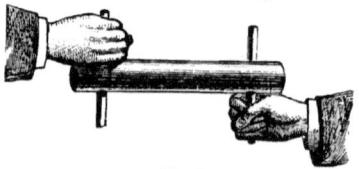

Fig. 5..

I will next try to stretch it out by hanging it up
tightly by one end, and then applying to the other
end a heavy weight, but it won't be stretched.

I will now, by means of two rods, fitting into the
bar at its ends, as you see in the figure, try to twist
round the one end, while I hold the other still, but
it won't be twisted.

I will now set the bar endwise upon the table, and put a heavy weight above it, to try and squeeze it together, but it won't be squeezed.

And finally I will hang it up horizontally by both ends, and attach a weight to the centre, and I find it won't be bent.

Now the bar of iron which I can neither break by a blow, nor stretch, nor twist, nor squeeze together, nor bend, is a very good example of a solid body ; and yet, if I applied an exceedingly great force, this bar might be stretched, or twisted, or squeezed, or bent. And in truth I did actually stretch, and twist, and squeeze down, and bend it, in the experiments I have just described, but not enough to make it visible to you. In fact the amount by which I stretch, or twist, or squeeze down, or bend the bar, depends upon the amount of force I use ; and in Physics we try to find out the relation between the force which we use and the effects which we produce. I cannot tell you all about this subject, because it would take up a great deal of time, but we may take one operation, such as bending, and endeavour to find in what way its effects depend upon the force which we employ.

15. Bending. EXPERIMENT 9.—For this purpose let us support a wooden beam in a horizontal position by both ends, and let us hang a somewhat heavy weight from its middle or centre. Then let us measure upon a scale how far the centre has been bent down by the weight. Let us now double the weight that hangs from the centre, and mark the new position

of the centre of the beam under the increase of weight,
and we shall find that the centre of the beam has been
lowered about twice as much by the double weight as

Fig. 6.

by the single weight, or in fact
the bending is nearly proportional to the weight applied.

EXPERIMENT 10.—Let us now
take the very same beam of
wood, and place it in edgewise,
so as to give it a great depth,
rather than a great flat surface,
and let us apply the same force
as before. We shall find that the beam is not bent
nearly so much as it was before.

16. **Strength of Materials.**—Now if an architect or an engineer were using great wooden beams in
the construction of a building, it would evidently be
most advantageous to strength were he to place them
in such a way that their depth might be as great as
possible, for in such a position they would give way
much less under any heavy weight.

An architect or engineer ought therefore to know
all about the strength of things, and how to place
them so as to get the greatest possible strength out of
the least possible amount of material ; in fact he
ought to know how to use his wood or his iron in the
best possible way.

Another point that the architect or engineer should
bear in mind is to make his house or his bridge five
or six times strong enough to bear the greatest load

that will ever be put upon it.   For sometimes a building may be strong enough to stand a heavy weight
on the floor, or a bridge may be strong enough to
stand the passage of a long train, without absolutely
breaking down, and yet the floor of the building may
be so much bent that it won't quite recover itself
when the weight is taken off, or the bridge may in
like manner be so much bent that it won't recover
itself when the train has passed.   In such a case the
floor will be less strong each time the weight is put on
it, and the bridge will be less strong each time the train
passes.   They will in fact go on bending more and more,
until at last they give way.   The architect or engineer
must therefore take great care that his structure is
never bent beyond the **limits of perfect recovery.**

17. **Friction.**—Before leaving solids, let us say a
few words about **friction.**   If I put a very heavy
weight on the table, it will require a very strong force
to move it along.   But if the table were of marble
and not wood, then a much less force would make the
weight slide along, while if the weight were on a sheet
of ice it would move with a still smaller force.   Now
the force which makes it difficult for me to push along
a heavy weight, is called the force of friction.

We should fare almost as badly without friction as
we should without the other forces : for if there were
no friction, we should be always walking, as it were
on ice ; and if there were the slightest slope, nothing
would be able to stand upon it, but everything would
slide down to the bottom.

PROPERTIES OF LIQUIDS.

18. **They keep their Size.**—In a liquid such as water, we can move the particles about very easily, but we cannot by any means force a quantity of water into smaller size, or make a quart content itself with a pint bottle.

* EXPERIMENT 11.—Let us, however, try to do so, and see what result we get, because we ought always to make an experiment when we can. Let us take a quantity of water shut in at one end, while at the other there is a water-tight piston or plug. Now let us try to drive this piston down in order to force the water into smaller volume, and to do so let us put a large weight upon the piston ; but notwithstanding all this we cannot compress the water.

19. **They communicate Pressure.** EXPERIMENT 12.—Let us now take a quantity of water shut in by two plugs or pistons. If we push the one piston down, we cause the other to mount up. Now if we put a ten-pound weight on the one piston, and an equal weight on the other piston, the one will exactly balance the other, and neither will be moved.

* EXPERIMENT 13.—In the last experiment both pistons were vertical, as in Fig. 7 ; but now let the one piston be vertical and the other horizontal, and by means of a simple arrangement apply a ten-pound weight to the horizontal piston. If now we apply a ten-pound weight to the vertical piston, we shall exactly balance the ten-pound weight attached to the horizontal piston. If, however, we apply a twelve-pound weight to the vertical piston, we drive along the horizontal piston; and in like manner if we apply a twelve-pound weight to the horizontal piston, we drive up the vertical piston. Thus, by means of the water we can convert the downward push of the ten pounds on a vertical piston into an equal push, only horizontal and outwards against the other piston. And thus you see a liquid such as water communicates pressure in all directions. This fact was found out by Pascal.

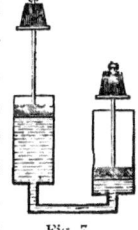

Fig. 7.

* EXPERIMENT 14.—In this experiment we have two vertical pistons, but the surface of the one piston is double that of the other. Now if we put ten pounds on the smaller piston, it will no longer be balanced by the ten pounds on the larger piston, but we shall require to put twenty pounds on the larger piston, in order to balance the ten pounds on the smaller piston. In like manner, if the large piston has three times the surface or area of the small one, we shall find that ten pounds on the small one will balance thirty pounds on the large one. Not only, therefore, does the downward pressure on the one piston communicate an upward pressure to the other, but the whole upward pressure is proportional to the surface of the piston; so that if the one piston has three times the surface of the other, it will be

B

driven up with a pressure three times as great, and so on.

20. **Water Press.**—Now this is a very valuable property of water, and it has been made use of in the construction of a very powerful machine, called the Bramah Press, from the name of its inventor. We have here a figure of it. You see a couple of bales of wool which we wish to squeeze as much together as possible, in order that they may occupy little space

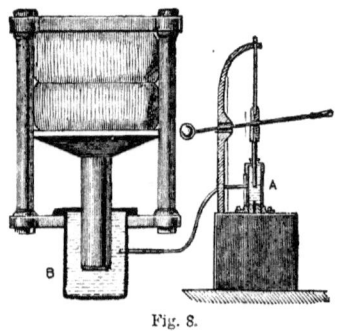

Fig. 8.

when carried about from one place or country to another. You see also two pistons—a large and a small piston—the large piston having one hundred times the area or surface of the small one. Now if I put a ton on the small piston, I must put a much greater weight on the large piston to keep it down, for the large piston is one hundred times the area of the small one. I must therefore put one hundred tons on the large piston in order to balance the ton

on the small piston, so that this large piston will rise
with the enormous force of one hundred tons, and
press with this force against the bales of wool, which
will therefore be squeezed very tightly together. It
is necessary, of course, in a machine of this kind, that
every part of it should be very strong and very
tight, otherwise the water would burst out with
immense force through any crevice or weak part.

21. **Liquids find their level.**—The next property
of liquids is that they always place themselves so as
to have a level surface. You will see at once that
this surface could not be slanting, for then the part
which is high up, having no friction, would slide
down towards the lowest part. A geometrician
would tell us that if we hang a plumbline above a
surface of water, this plumbline will be perpendicular
to the surface; that is to say, it will not slant
towards the surface in any one direction, but will
stand straight up, and we may show this by a very
simple experiment.

EXPERIMENT 15.—Take all the mercury in the
bottle and pour it into a flat vessel, and get it to
cover all the bottom of the vessel by making the
vessel level. Now hang a plumbline over the vessel,
and you will see that the reflexion of the plumbline and the plumbline itself are in one direction,
so that the one appears to be a continuation of
the other. This shows that the plumbline does not
slant towards the surface; for if it did, the reflexion and the plumbline itself would not form one

line, but would appear as two lines bent towards one another.

EXPERIMENT 16.—Even when the liquid is contained in bent tubes, that in the left-hand tube will always be at the same level as that in the right, and this will take place whatever be the shape of the tube. Indeed, I have only to fill some of these curiously shaped tubes with water in order to convince you that this is the case. You see the water is at the same level in all the tubes.

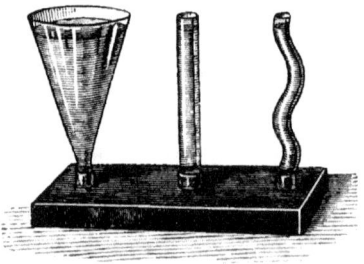

Fig. 9.

22. **Water-level.**—And this leads me to speak of the water-level which you see in the figure. If I place my eye in a line with the top of the water in both the ends of the tube, I know that I am looking along a level line, and that all the points near me which I see along this line are precisely at the same level, so that if a flood were to come it would reach them all precisely at the same moment.

It is often very important to know what points are

on the same level : a man who constructs a canal or
a railway, must know this ; and in order to do so, he

Fig. 10.

must use a level of some kind.   The kind which is
most often used is called the **spirit-level**; that which
we have described is called the **water-level.**

23. **Pressure of deep Water.**—Let us now take
a somewhat deep vessel filled with water.   You will
see at once that the layers of water near the bottom
are pressed upon by the weight of all the water above
them, so that the pressure upon these layers will be
greater the further they are below the surface.   In
fact, the layers two feet below the surface will be
pressed upon with twice as much water as those only
one foot below ; in other words, the **pressure will
be proportional to the depth.**

EXPERIMENT 17.—This pressure will act in all
directions, upwards and sideways, as well as downwards.   To show this let me nearly fill a vessel with
water and withdraw a plug from the side near the
top.   You see the water is pushed out by the pressure
upon it, but not very forcibly ; let me now withdraw
a plug near the bottom, and you see that, owing to

the great weight of water above, the pressure is now
much stronger, and the water rushes out with great
force. So much for a pressure sideways. I shall
now try to show you that there is also an upward
pressure. To do so I take what is called a cylinder
or wide tube of glass without either top or bottom.
But here you see I have a separate closely fitting
bottom which I attach to it, and you see, too, that

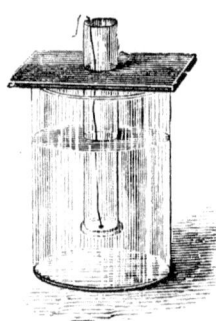

Fig. 11.

I have a string coming up
through the cylinder, by
which I can hold it tightly
on. Holding it on by the
string, I will now plunge the
cylinder below the surface of
water in the vessel, and you
see that I may now let go
the string, but yet the bottom
does not fall off because it
is kept on by the upward
pressure of the water against
it. I will now pour a quan-
tity of water coloured blue by indigo into the cylinder,
and yet the bottom is held on, and it will only drop
off when the water in the inside of the cylinder has
reached to nearly the level of the water on the outside,
because then the upward pressure against the outside
of the loose bottom is balanced by an equal downward pressure of the coloured water against the
inside of the same.

If any of you should ever be in a boat on deep

water, you may easily prove to yourselves the great
pressure of water at a great depth. Take an ordinary
quart bottle and fill it three-fourths full of water;
then cork it tightly, and attaching it to a long string,
let it down into the deep water. If it be allowed to
descend sufficiently far, the pressure of the outside
water will be so great as to force the cork into the
bottle, and when you pull it up you will find the
bottle full of water with the cork inside.

24. **Buoyancy of Water.**—Let us now try to
get precise ideas about the buoyancy, or floating
power of water; and, to do this, let us make one or
two experiments.

EXPERIMENT 18.—Let us take our balance, which
we have previously spoken about (page 23), and
get it into order for weighing. Now here we have
a substance which weighs 1,000 grains, as you see,
when we make the weighing in air. Let us now
attach the substance to the right hand scale-pan, and
make the weighing in water. What is the result?
We find that actually it appears to have no weight
at all, and I require to put on the right-hand scale-
pan 1,000 grains, or the whole weight of the substance,
in order to make it equal to the other scale-pan in
weight.

EXPERIMENT 19.—Are we to imagine that this substance, when in water, loses its weight altogether?
Let us try, by experiment, whether or not this is the
case. First of all I shall place a vessel with some
water in it on one scale-pan, and balance it by weights

in the other. I now drop the substance weighing 1,000 grains into the water, and you see the result. The scale-pan with the water having the substance in it is now much too heavy, and I have to put 1,000 grains into the other in order to restore the balance.

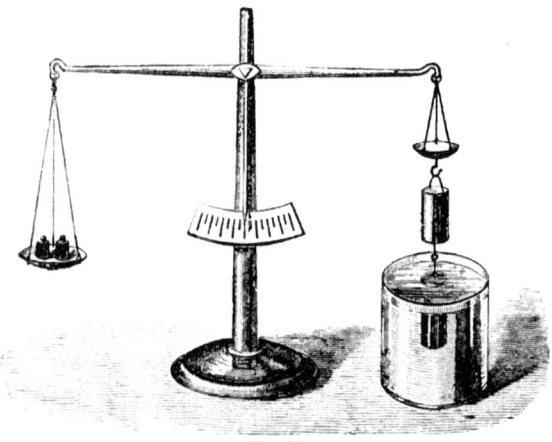

Fig. 12.

But this is precisely the weight of the substance, and therefore you see the substance does not really lose its weight. The weight is still there; that is to say, the vessel with the substance in it is 1,000 grains heavier than if the substance were not there, but the substance itself has its weight apparently taken away by the buoyancy of the water, which acts as an upward pressure.

EXPERIMENT 20.—Here we have (fig. 12) a brass cylinder which fits, as you see, exactly into a hollow socket. Let us now take it out of the socket and attach it, as well as the socket, to the hook at the bottom of the right-hand scale-pan (see figure), and let us counterpoise them both so that they are exactly balanced. Let us now weigh the cylinder, not in air, but water, by placing a vessel containing water below the right-hand scale-pan so that the cylinder is wholly immersed in the water. The right-hand scale-pan is now too light. The brass cylinder has, in fact, lost part, though not all, of its weight, by being weighed in water. To see how much, we will pour some water into the empty socket which is hung below the scale-pan. Now we have exactly filled it with water, and we have, at the same time, restored the weight which the brass cylinder lost through being weighed in water, for now you see the two scale-pans are balanced once more. But the brass cylinder exactly fitted into the socket, so that we have added water exactly equal in bulk to the brass cylinder (that is to say, a socketful) in order to restore the loss of weight. We gather from this that the brass cylinder, when weighed in water, appeared to suffer a loss of weight exactly equal to the weight of its own bulk of water, and we may extend this to any other substance, and say that **when anything is weighed in water it will suffer a loss of weight exactly equal to the weight of its own bulk of water.**

25. **Flotation in Water.**—Let us now see what

this means. It means that if a substance immersed
in water be heavier, bulk for bulk, than water, such
as the cylinder, it will suffer a loss of weight equal to
the weight of its own bulk of water, but yet it will
not appear to lose all its weight, because it is heavier,
bulk for bulk, than water is.   It will therefore fall to
the bottom because it will still have weight.

EXPERIMENT 21.—If, however, the substance be
of the same weight, bulk for bulk, as water, such as
that of Experiment 18, then it will lose all its weight
when in water, and will not sink. If I therefore
put this substance into water, you see it neither sinks
nor swims, but moves about anywhere, just as if it
had no weight.

Now what will happen if the substance be lighter,
bulk for bulk, than water?   How can it lose more
than its own weight? you may ask.   Let us learn, by
means of experiment, what will take place in such a
case as this.

EXPERIMENT 22.—Here I have a piece of wood
which is lighter, bulk for bulk, than water, and I force
it beneath the surface of the water ; but I find that
the upward pressure caused by the buoyancy of the
water is now greater than the weight of the substance,
so that it is forced up to the top of the water and
swims upon the surface.

Well, as the result of all these experiments, we may
conclude, firstly, that any substance immersed in water
appears to become lighter by the weight of its own
bulk or volume of water.   And secondly, that in

consequence of this, if the substance be heavier, bulk for bulk, than water, it will sink ; if of the same weight, bulk for bulk, as water, it will neither sink nor swim ; but if lighter, bulk for bulk, than water, then it will swim.

26. **Comparative Density.**—Now I wish to show you that we have here got a method by which we can tell how much any substance is heavier, bulk for bulk, than water.

* EXPERIMENT 23.—Let us imagine that we have a small piece of gold that weighs in air exactly 19 grains—that is its weight. Let us next weigh it in water, and we find that it now weighs only 18 grains, showing a loss of weight equal to 1 grain. Now this loss is equal to the weight of its own bulk of water, which is therefore 1 grain. But the gold in itself weighs 19 grains, so that it weighs 19 times as much as its own bulk of water. This is what we mean when we say that the **specific gravity** of gold is 19. Now we shall get the same result whatever be the size or shape of the piece of gold we use. But on the other hand, if a person put something into our hand that was not really gold, but only like it, we should no doubt find by weighing it in water that the substance was not so much as 19 times heavier than its own bulk of water. This method of finding out the specific gravity or relative density of bodies was discovered more than 2,000 years ago by a philosopher called Archimedes. Hiero, King of Syracuse, had a crown of gold, and he had reason to believe that the goldsmith had mixed a quantity of silver with the gold, but he could not think of any way of finding this out—so in his difficulty he applied to Archimedes. The true way of finding it out occurred to Archimedes one day when he had gone to take a bath, and the tradition is that he immediately ran out of the bath quite naked, shouting out " Eureka ! Eureka ! " which means " I have found it out ! I have found it out ! " He then went home and got a piece of gold which he knew was pure, and found that when weighed in water it lost one-nineteenth part of its whole weight, from which he argued as we have done, that pure gold is nineteen times as heavy as water, bulk for bulk. He next took Hiero's crown, but he found that when weighed in water it lost more than one-nineteenth part of its whole weight, from which he argued that it was not made of pure gold, and doubtless the goldsmith was properly punished for his theft.

27. **Buoyancy of other Liquids.**—Other liquids besides water have buoyancy. Indeed, each liquid has its own peculiar amount of buoyancy. A very light liquid, such as alcohol or ether, has comparatively little; while a very heavy liquid, such as mercury, has a great deal. To convince you of this, I have only to pour some of this mercury into a vessel, and put on its surface a bit of iron—the iron, as you see, floats; showing that it is lighter, bulk for bulk, than mercury. Gold, on the other hand, is heavier than mercury; in fact, mercury is $13\frac{1}{2}$ times as heavy as water, bulk for bulk ; while gold, you have

already seen, is about 19 times as heavy, bulk for bulk.

Salt water is somewhat heavier than fresh; and there is in Palestine an inland lake called the Dead Sea, so salt, and consequently so heavy, that a man immersed in it could not possibly sink.

28. **Capillarity.**—Before leaving liquids, let me just mention a well-known case in which water will rise above its own level.

EXPERIMENT 24.—If we hold a lump of sugar above the surface of water in a vessel, and allow its lower end to touch the surface, we shall soon find the whole lump wet. In like manner, if we dip a strip of blotting-paper or cotton-wick in water, we may convey it above its level by these means.

But if we hold the sugar or strip of blotting-paper with its lower end touching a surface of mercury, the mercury will not rise into the sugar or the blotting-paper; so that these two liquids, water and mercury, behave differently as regards the lump of sugar or the strip of blotting-paper. In the first place, we see the water rise into them, and not only rise into them, but remain there; the mercury, on the other hand, will not rise into them and will not wet them; in fact, mercury has not a sufficient attraction for sugar to rise into it, nevertheless mercury may be made to adhere to a surface of silver or of gold, because it has a great attraction for these metals.

PROPERTIES OF GASES.

29. Pressure of Air.—Gases have many points
of likeness to liquids, but in other respects the two
are very different. A liquid has a surface, so that you
may fill a bottle half full with a liquid and shake the
liquid against the sides of the bottle. But you cannot do this with a gas. Here, for instance, I have a
bladder which contains gas, but the gas fills the
whole bladder, and not a part of it. In fact, a gas
has an intense desire to fill any vacant space that is
not already filled, and will strongly exert itself to
do so.

EXPERIMENT 25.—I can easily prove this by a very
simple experiment. I have here an air-pump which
I will afterwards describe to you ; meanwhile let me
tell you that by means of this air-pump, we can take
out of this bell-jar the atmospheric air which it now
contains. You see the india-rubber ball full of air
which I will put under the bell-jar. Now I will
exhaust the bell-jar, that is to say, take its air out,
and what is the result ? There is air in the india-
rubber ball, but there is now none round about it,
and in consequence the air in the ball tries to fill
the empty space, but it can only do this by enlarging
the ball, and you see the ball grow bigger and bigger
as I continue the exhaustion. I shall now let the
air in, and you see the ball once more resumes its
former size.

EXPERIMENT 26.—We may vary the experiment in this way. I shall now place on the bed-plate of the air-pump a jar which is covered at its top by a piece of india-rubber tied tightly round the rim. I now exhaust the jar as before, and find that as I withdraw the air from the inside of the jar, the outside air trying to force itself into the void space presses down the india-rubber cover, and per-

Fig. 13.

haps, before the experiment is over, the pressure may be great enough to burst the india-rubber.

30. **Weight of Air.**—You thus see that air will force itself into any space that is empty, if it possibly can, and indeed we have the greatest difficulty in emptying all the air out of any vessel. We can, however, take out the greater part of the air which fills a vessel. In fig. 14, for instance, is a vessel which we can attach to the air-pump, and by this means deprive it of air, and it will be found that the vessel full of air weighs heavier than the vessel empty, or, in other words, **air has weight.**

Fig. 14.

EXPERIMENT 27.—Let us now attach a light box *bottom downwards* to one of the arms of the balance, and ascertain its weight. This weight may be said to be that of the box filled with atmospheric air.

EXPERIMENT 28.—While this light box remains counterpoised let us next fill it by displacement (see Instructions) with a heavy gas called carbonic acid gas, which you are told how to make in the Chemistry Primer, Art. 33. You see that the pointer is displaced, showing that the vessel now weighs heavier than when it was filled with ordinary air, so that some gases are heavier than others.

EXPERIMENT 29.—Hydrogen is the lightest of all gases, and accordingly let us now attach the box *bottom upwards* to the arm, and when it is counterpoised fill it by displacement (see Instructions) with hydrogen, which you are told how to make in the Chemistry Primer, Art. 17; the pointer will now be displaced in the opposite direction, showing that the vessel weighs much lighter than when filled with air, although not so light as if it had nothing in it. We learn from this, that although the particles of gases appear to repel each other, trying to get as far from one another as they possibly can, and always filling the vessel that holds them, yet they are attracted by the earth and have weight, so that there is no danger of our atmosphere rushing away from the earth. Instead of this the atmosphere clings to the earth as a sort of ocean, and at the bottom of this ocean of air we all live and move.

Now as far as regards pressure and weight an ocean of air is similar to one of water, and you may remember you were told, page 37, that the pressure of water against the bottom of a vessel depends upon its depth, so that at a great depth you have a great pressure, and this pressure is exerted in all directions.

Now if you are told that we have a great pressure of air upon us, you will naturally ask,—How is it then that we do not feel this pressure? We reply— simply because the pressure is exerted in all directions, upwards, downwards, and sideways. Take a sheet of paper—the pressure of the air not only acts on the top of the sheet pressing it down, but it acts just as strongly on the bottom of the sheet pressing it up, and in consequence the sheet of paper can move about freely just as if there was no pressure of the atmospheric ocean upon it at all. And for the very same reason you and I move about freely and do not feel the pressure. Notwithstanding, I hope to convince you by a simple experiment that we can make the pressure of the air very perceptible.

EXPERIMENT 30.—Here are two hollow half-spheres which exactly fit on to one another. Now let us press them together and shut the stopcock, and you will naturally ask why does not the pressure of the air hold them firmly together? The reason is that there is also air within them, and this air presses outwards just as much as the air without them presses inwards. But now let us fit on these two half-spheres to the air-pump and take the air out of them, and having done so let us shut the stopcock, and detach them from the pump ; you will now find it very difficult to pull the two half-spheres asunder, because while the

air from without presses them together there is no air
from within to counteract this pressure, and they are '
in consequence held very firmly together.

Now, since air is a fluid, and has weight, it will
have a certain amount of buoyancy, although not
nearly so much as water.   If, therefore, a large bag

Fig. 15.

be filled with coal gas, or, better still, with hydrogen,
it will be lighter, bulk for bulk, than the air, and
will therefore rise in it. Such a bag is called a
balloon, and if sufficiently large it may also support
a small car containing several people.

31. **Barometer.** EXPERIMENT 31.—Let us now
take a hollow tube of glass, open at one end and

closed at the other, fill it with mercury, and keeping
the finger tightly against the open end invert it into
a glass vessel also containing mercury, taking care not

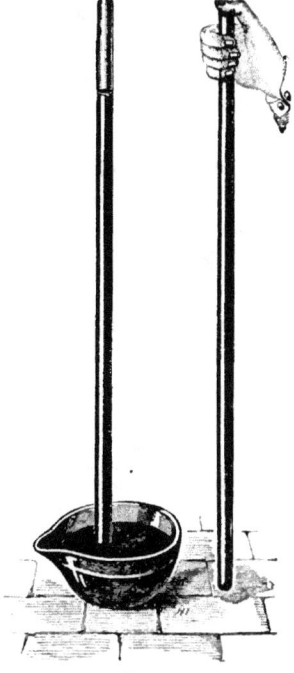

Fig. 16.

to withdraw the finger from the open end until this
end is below the surface of the mercury in the glass
vessel.   Here you see (fig. 16) we have the tube so

inverted standing upright in the vessel of mercury.
Now mark what happens. You see a blank space left
at the top of the upright tube of mercury, and your
first idea is that we must have let some air in, but
this is not the case. There is absolutely nothing in
this blank space. You are next inclined to ask,
Why does not the atmospheric air, which is no doubt
pressing in all directions, and therefore pressing upon
the surface of the mercury in the vessel, drive up the
mercury so as to fill this empty space? The reply is
that it would if it could; as it is it presses upwards
against the surface of the mercury in the vessel with
force sufficient to keep up in the tube a column of
heavy mercury thirty inches high; but it can do no
more—the weight of this mercury pressing downwards exactly counterbalances the pressure of the air
forcing it upwards, and hence on the one hand the
column of mercury cannot push itself downwards,
and on the other the pressure of air cannot push
the column upwards, and we have therefore a blank
space above the column. This experiment was devised by an Italian called Torricelli—the tube is
called a **Barometer,** and the empty space at the top
is called the **Torricellian Vacuum.** Most baro-
meters are provided with a scale of inches by which
the height of the top of the column above the surface
of mercury in the cistern may be accurately measured.

32. **Uses of the Barometer.**—The barometer is
useful in many ways; for instance, we may by its
means tell the height of a mountain. You were told

(page 37) that the pressure is greater at the bottom of a deep vessel of water than near the top, and the same thing takes place in this ocean of air in which we live— the pressure is greater near the bottom of this aërial ocean than it is far up near the top. If therefore we go to the top of a high mountain, we have a smaller weight of air above us than we had when down below, and in consequence the pressure of the air will be smaller at the top of the mountain than at the bottom. The air will not now be able to balance the same column of mercury as at the bottom, so that, in the barometer, instead of a column of mercury thirty inches high, we shall only have one of twenty-five inches or possibly of twenty inches, depending upon the height of the mountain. In fact the mercury will sink lower and lower down in the tube of the barometer the higher up you rise in the air, and thus by means of the barometer you can tell to what height you have gone. The barometer is also useful in telling us when bad weather is at hand. When the barometer falls, that is to say, when the top of the column of mercury gets lower in the tube and especially when it falls quickly, we may expect bad weather. On the other hand, if the mercury remains steady and high we may expect a continuance of fine weather.

33. **Air-pump.**—We have already spoken about taking the air out of a jar, now this is done by the **air-pump.** You will see how this instrument acts by means of the figure. But first of all I must tell

you what is meant by a **valve.** A valve is just a
tightly fitting trap-door that closes a hole, and that
can only open in one way—upwards, for instance.
You have, most of you, seen trap-doors in floors that
open upwards. Now in the figure you see to the left
a bell-jar full of air, which fits tightly upon a plate.
You see too coming out from the middle of the plate
a tube which opens into the bell-jar on the left side,
and into the cylinder or barrel on the right, and thus
connects the two together. You see also a piston

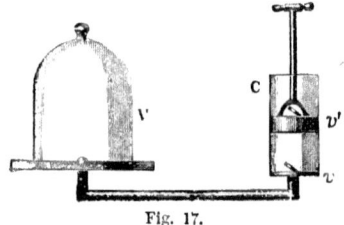

Fig. 17.

or plug that can move up and down in the cylinder
or barrel.

Finally you see two valves or small and tightly
fitting trap-doors, one of which is placed where the
tube enters the bottom of the cylinder, while the
other is in the piston itself. Both of these valves
open upwards and not downwards.

Now suppose we start with the piston at the
bottom of the cylinder, and the valves shut, and
begin to pull the piston up. In doing so we make an
empty space which the air on all sides will try to fill
up if it possibly can (Art. 29). The air from above

will try to press into this space, but it will not be able to get in, and all it can do will be to press against the outside of the upper valve and keep it tightly shut, since the valve does not open downwards. The air from the bell-jar will succeed better, for it will rush through the tube and press open the lower valve which opens upwards, and then get into the empty space. Let us now suppose that we have got the piston to the top of the cylinder, and that we are beginning to press it down. The push that we give to the piston, the piston gives to the air; and the air in its turn communicates this push to the lower valve, which is kept shut. But the air within is more successful with the upper valve, for it pushes this open; and so, as we continue to push down the piston, all the air that was in the cylinder below it is pushed out through the upper valve or trap-door. But this air which we have pushed out was part of the air that was originally in the bell-jar, so you see that in the first double or up-and-down stroke of the piston we have succeeded in squeezing out part of the air of the jar. Let us now repeat the same process, that is to say, raise the piston again, and the air from above will shut the upper valve, while the air from the bell-jar will rush along the tube, push open the lower valve and fill the empty space which we make when raising the piston. And when the piston descends once more, the lower valve is kept shut, while the air within pushes open the upper valve and gets out, and thus in every double stroke we get rid of part of the air in the bell-jar. Of course it is quite necessary in working the pump that the piston shall fit quite tightly into the cylinder ; for, if not, the air will get in from without, and therefore we shall not succeed in getting the air out from within. I have now told you the way in which the air-pump works, but you must not expect every air-pump to be precisely like the figure I have given you ; the principle, however, of all air-pumps is the same, although the appearance may be very different in each.

34. **Water-pump.**—Having now told you about the air-pump, let us return for a moment to the barometer. You have seen how the pressure of air is just strong enough to hold up a column of mercury about thirty inches high. But water is much lighter, bulk for bulk, than mercury, and we might therefore expect the pressure of the air to hold up a much longer column of water than one of thirty inches. In truth, the pressure of the air will hold up a column of water very nearly thirty feet in height.

This will enable you to understand the mode of action of the common pump. In the figure on the next page you have a sketch revealing the interior of such a pump. Below we have the reservoir from which we wish to pump the water up, and we have a tube leading from this reservoir up into the barrel of the pump. In this barrel you see a piston which fits tightly into the barrel, and in this piston there is a valve opening upwards, while at the bottom of the barrel there is another valve also opening upwards.

In fact, the barrel of the lifting pump is quite similar to that of the air-pump, and we may begin by supposing that the piston is at the bottom of the cylinder. Let us now raise up the piston, and just as in the air-pump, the air above will press down the upper valve and keep it shut. The air in the tube will on the other hand rush up through the lower valve in order to fill up the empty space made by raising up the piston. When we lower the piston again, just as in the air-pump, the lower valve will be shut, and the valve in the piston will open and let out some air. In fact, we are now pumping out the air from the barrel and the tube. But meanwhile, what is the water in the reservoir doing? The air from without continues pressing on the surface of the water in the reservoir; but as we have been taking away the air in the tube, this pressure of outer air is no longer counterbalanced by that of the air in the tube; the outer air will therefore find itself unopposed, and will drive up the water into the tube, until at last, when all the air is taken away, the whole tube will be filled with water. This water will then enter the pump-barrel through the lower valve.

Fig. 18.

But all this will not take place if the distance between the surface of water in the reservoir and the lower valve be more than thirty feet. For you have just been told that the pressure of the air will support a column of water thirty feet high, but if the column be higher than this it will not support it. So that if there be a greater distance than thirty feet between the surface of the reservoir and the pump barrel, the water will refuse to enter into the barrel, and do what you can you will not be able to entice the water quite up into the barrel. If, however, the distance be not more than about twenty-six or twenty-seven feet, the pump will work well, and you will get the water to enter the barrel. Suppose now that you have got the barrel filled with water, and that you are pressing down the piston. As you do this the pressure you give the piston will be communicated by the water to the lower valve, which will be kept closed. On the other hand, the pressure of the water will force open the upper valve which opens upwards, and the water will get above the piston. Next time when you pull up the piston, you will pull up this water with it, and it will empty itself through the spout of the pump, and the water will now come out of the spout at every stroke.

  * EXPERIMENT 32.—To enable you to see with your own eyes what goes on in a common pump, take a model in which the pump barrel is made of glass, so that you can see into it. You will thus see that when we raise the piston, the upper valve shuts and the under one opens, while, as the piston descends, the under valve shuts, and the upper valve opens. You quite understand that the piston of the pump must fit tightly onto the barrel, because otherwise the air will get in from above and prevent the action. Sometimes, however, if a pump is not much used, the leather or other packing around the piston gets dry, and the pump will not act. In that case, if a little water is thrown upon the piston, it wets the packing and serves to make it tight.

35. **Syphon.**—Before leaving this subject, let me describe to you an instrument called a syphon, of

Fig. 19.

which the action depends, like the pump, upon the pressure of the air. I shall not, however, explain its principle. You see the syphon before you in the figure ; it is used for conveying liquids from a vessel at a higher to one at a lower level.   In the first place, you must invert the syphon tube, and completely fill it with water, keeping your finger at the end of the shorter tube.   Now place the shorter end beneath the surface of the water in the higher vessel as in the figure, and remove your finger.   Once you have done this, the water will, thereupon, flow in a continuous stream from the end of the longer tube into the lower vessel, and you may by this means remove the water completely from the upper into the lower vessel, provided the short tube of the syphon be long enough to reach to the bottom of the upper vessel.

MOVING BODIES.

36. **Energy.**—You have been told (page 9) about the moods or affections of things, and how a cannon-ball in motion is a very different thing from one at rest, or a hot cannon-ball from a cold one ; and you have also been told that one of our great objects in this Primer is to find out something about these varying moods or affections of matter.   We could not begin with this, for we had first of all to tell you about the things themselves, and you ought now to have a tolerably good acquaintance with solids, liquids, and gases ; it is time, therefore, that you learned something about the varying moods or affections of things.   You were told that bodies were sometimes

full of energy, such as a cannon-ball in motion, and
sometimes utterly listless and devoid of energy, such
as a cannon-ball at rest, and in what follows we cannot do better than study the most conspicuous cases
in which a body is full of energy. Now this happens
when a body is in **actual motion**, or when it is in
rapid vibration, or when it is **heated**, or when it is
electrified, and we shall therefore class energetic
bodies under these four divisions. We shall first of
all speak of bodies in actual motion, and under this
head give you some idea of the way of acting of such
bodies ; we shall then speak of bodies in vibration,
such as a sounding drum or bell, and under this head
we shall tell you something about sound. We shall
next speak about heated bodies, and under this head
tell you something about light and heat ; and lastly,
when speaking about electrified bodies, you will
hear about that mysterious thing called electricity.
We cannot in this little Primer give you anything
like a complete account of the various moods of bodies
or the various kinds of energy which they sometimes
possess. This must be reserved for a more advanced
stage ; we can only give you a mere outline of the
subject, telling you at the same time that it is one of
very great importance.

37. **Definition of Work.**—When we say that a
man is full of energy, we mean that he is full of the
power of doing work ; and when we say that a thing
is full of energy, we mean in like manner that it is
full of the power of doing work. In fact, we measure

the energy of anything by the amount of work
which it can do before it is utterly spent. ˙ Now if we
raise a pound weight one foot high, we do a certain
amount of work, but if we raise it two feet high we
do twice as much work, if three feet high three times
as much work, and so on. If therefore we call the
work of raising a pound weight one foot high one,
we should call the work of raising it three feet high
three.

Again, the work of raising two pounds to any
height is double the work of raising one pound to the
same height, so that the work of raising two pounds
three feet high would be six. In fact, multiply the
number of pounds you raise by the number of
feet you raise them, and the product will give
you the work done.

Let us now suppose that we point a cannon straight
up into the air, and discharge a ball weighing 100 lbs.
with velocity just enough to make it mount up
1,000 feet before it turns ; we can tell at once from
this how much energy the ball had when it was
discharged. It had energy enough to carry 100 lbs.
(that is to say, itself) up 1,000 feet, and consequently
energy enough to do work equal to $100 \times 1,000$ or
100,000. If we now put a larger charge of powder
into the cannon, we shall make the ball come out
with greater velocity. Suppose that now it can
mount up 1,500 feet before it turns ; it has therefore
energy capable of doing work equal to $100 \times 1,500 =$
150,000. In fact, you see at once that the greater

the velocity or quickness with which the ball is shot out, the higher will it go, the more work will it do, and hence the greater energy will it have.

38. **Work done by a moving body.**—I cannot enter very fully here into the subject, but I will tell you that a body shot upwards with a double velocity will mount not **twice** but **four times** as high—a body with a triple velocity not **thrice** but thrice three times or **nine** times as high—and so on.

You see therefore that a cannon-ball of double the velocity will do four times the work. But there are other ways of measuring the work of a cannon-ball than by seeing how high it can lift itself into the air, for we may fire it into wooden planks placed one behind the other, and we shall then find that a ball with a double velocity will go through nearly four times as many planks, a ball with a triple velocity through nearly nine times as many, and so on. You thus see that a ball with a double velocity will have four times the destructive effect of one with a single velocity, and indeed in whatever way we measure its energy it will have four times as much energy as the other.

39. **Energy in repose.**—It is very easy to see that a body moving very fast has the power of doing a great deal of work, but besides this we have often energy in a quiet state, just as a man may be quiet and yet able to do a great amount of work when he sets about it. Suppose there are two equally strong men fighting together, each with a heap of stones

which they are throwing at each other, only the one
with his pile of stones is standing on the top of a
house, while the other man is standing at the bottom
with his pile.   I need not ask you which of the two
is likely to win the day ; you will tell me at once the
man at the top of the house.   Now why has he the
advantage ?   He is not stronger or more energetic
than the other—his advantage is therefore due to the
stones ; it is clearly because his pile of stones is high
up.   He himself has not more energy than the man
at the bottom, but his pile of stones has more energy
than the pile of stones of the man at the bottom, and
thus you see that the stones have an energy arising
from the high position in which they are placed ; they
are, in fact, capable of doing work, whether this be
the very useless work of knocking down a man or the
very useful work of driving in a pile.   Or let us
suppose two water mills—one having a large tank or
pond of water at a high level near it, while the other
has a pond or tank of water, but at a level lower
down than that of the mill ; which mill is likely
to work ?   You will at once tell me, the one with the
pond of water at a high level, because the fall of
water will drive round the wheel.   You see, therefore, that there is a great deal of work to be got from
a pond of water high up, or a **head of water**, as this
is called—real substantial work, such as grinding
corn or threshing it, or turning wood or sawing it.
On the other hand, there is no work at all to be got
from a pond of water that is low down.

Let us now compare a water-mill driven by a head of water with a windmill driven by the wind. The wind is like the cannon-ball, although not moving so fast, its energy being that of a body which is actually moving : it is in fact rushing against the sails of the windmill and driving them round ; and if we throw up a feather or a straw in a strong gale, we find that it is hurried away by the wind. But a water-mill has one decided advantage over a windmill, for in a windmill we must wait for the wind ; but if we have a water-mill with a good head of water we may turn the water on and off whenever we choose. We can keep our stock of energy and draw upon it whenever we have a mind. In fact, the energy of a body in motion is like ready money which we are in the act of spending, but the energy of a head of water, or of any body which is high up, is like money in a bank, which we may draw out whenever we want it.

VIBRATING BODIES.

40. Sound.—A body that is changing its place is of course in motion, but it does not follow that every moving body changes its place as a whole ; a top that spins round very quickly is in motion, but it does not change its place as a whole.

EXPERIMENT 33.—Here is a wire which you see attached by one end to a support ; now if the other end be struck it goes backwards and forwards rapidly, but the wire as a whole does not change its place. When the particles of such a wire are moving backwards and forwards, they are said to be in a state of vibration. In like manner, when a bell or a drum is struck the particles of the bell or drum are in a state of vibration, or when the string of a musical instrument is pulled and let go, the string is in a state of vibration.

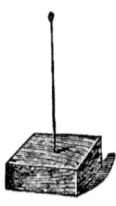

Fig. 20.

Now vibrating motion, just like motion from place to place, denotes energy, and indeed the particles of a vibrating body are moving actively about from side to side; if you try to stop them, they will give you a blow. If anything is in their way, they will give it a blow —the atmospheric air is, and they consequently give it a blow. Indeed each time the top of this vibrating wire comes back it gives the air a knock in the same direction. In fact, a vibrating body gives in a short time a great number of little knocks to the air. When the air is struck, it does not receive the stroke quietly, but strikes the air next it, and this in turn strikes the air next it, and so on, until the blow given to the air is carried over a great distance. At last this blow reaches your ear or mine, and we get a blow, which, however, does not affect us in the same way as a blow that knocks us down, and therefore we do not call it a blow; but we say that **a sound has struck our ears**—in fact we hear a sound.

41. **What is noise and what music ?**—Now if the body that strikes the air deals it an irregular series of blows, such as when a cannon is fired, the air carries this to our ear, and we say that we hear a **noise.** If however the body that strikes the air be in vibration, and deal it a great number of little blows in one second at regular intervals, the air will carry these on and give our ears just as many blows in one second at regular intervals, and then we say that we hear a **musical sound.** Thus you see that regularity or irregularity of interval constitutes the difference between a musical sound and a noise. More than this, if the vibrating body which is the cause of this disturbance deals the air only a comparatively small number of blows in one second, then the air will of course only deal us the same number in one second, and we shall hear a **deep low note ;** but if the vibrating body vibrates very quickly and deals the air a great number of blows in one second, the air will of course deal us just as many, and we shall hear a **shrill high** note. Thus you see a **deep low note means a small number of blows dealt to our ears in one second, while a shrill high note means a great number of blows in the same time.** A very shrill note will be given by 20,000 blows in one second, and a very low note by 50 blows in the same time.

42. **Sound can do work.**—A musical note is pleasant, but a noise is eminently disagreeable, and sometimes it hurts or even destroys the ear if

it be a very violent one. Thus if a large cannon were discharged, the blow to the ear might in some cases destroy its hearing power ; or if the sound struck against a pane of glass, the concussion might be so strong as to shatter the glass, and sometimes in such cases as the explosion of a powder magazine all the windows in the neighbourhood are shattered to pieces. Thus you see that a loud noise is something with energy in it, and that it can do work—more especially work of a destructive nature.

43. It requires Air to carry it. * EXPERIMENT 34.—Let us try to ring a bell in a place where there is no air, such as an exhausted receiver. There being no air, there will be nothing which the moving particles of the bell can give a blow to, and hence no sound will reach our ears. In fact, a bell that has been struck, or any other vibrating body, has in it a quantity of energy, of which it parts with some to the air, while the air in its turn parts with some to our ear. But if there be no air, there is nothing to carry to our ear the energy of the vibrating body.

44. Its mode of motion through Air.—Let us now think a little about the nature of this thing called sound, which is given out to the air by bodies in vibration, and which is then carried to a great distance by the air itself.

In the first place, when a cannon is discharged a mile or two off, do not imagine that the same particles of air travel all the way from the cannon to your ear.

The particles near the cannon give a blow to those next them and then stop, the particles that have received the blow give in their turn a blow to those next them and then stop, and so on, till the blow reaches your ear. What really happens will be made quite plain by the following experiment.

* EXPERIMENT 35.—Let us take a series of elastic balls suspended in a row by separate threads, so as to hang loosely together, just touching one another.

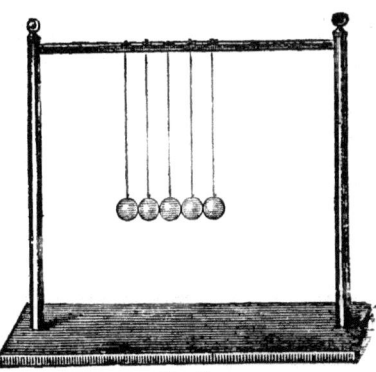

Fig. 21.

Let us now pull aside endways the first ball, and allow it to give a blow to the second. What will happen? The first ball having delivered its blow to the second, will become quite still. The second will very quickly transmit the blow to the third, and become still in its turn; the third will do so likewise, until the impulse reaches the last ball of the series, which being the

last will be put in motion by the blow. Now the
first ball may be likened to the particles of air which
are next the cannon, and the last ball to the particles
that are next your ear, and thus you see how the blow
from the air next the cannon is transmitted to the
air next your ear without the necessity of the same
individual particles of air moving all the distance in
order to carry it.

Those of you who have played at croquet must have
noticed what takes place when you croquet your
adversary's ball. In this case you hold your own
ball tightly under your foot while your adversary's
is just touching it : you then by means of the mallet
give a blow to your own ball, which does not however
move, but which transmits the blow to your adversary's
ball with sufficient force to send it a great way off.
We have here, therefore, a result the same as in the
series of balls.

45. Its rate of motion.—Again, this impulse or
blow which we call sound requires time in order to
pass from the cannon to our ear. No doubt it
travels very fast, as fast as a rifle-ball, but yet it
does not pass instantaneously from the cannon to
our ear.

Most of you have no doubt seen a cannon fired a
long distance off, and you then saw, first of all the
flash and puff of smoke, and after a few seconds you
heard the noise. Now these few seconds are the time
which the sound or impulse took to travel from the
cannon to your ear. You saw the flash the very

moment the cannon was fired, and therefore, counting from its appearance, you know how long the sound took to travel from the cannon to you. Suppose, for instance, that the cannon was 11,000 feet away, and that you reckoned ten seconds between the flash and the report, you therefore conclude that sound takes ten seconds to pass through 11,000 feet of air, or that it moves at the rate of 1,100 feet a second, which is pretty near the truth.

Sound will, however, pass through water much more quickly than through air, and by means of experiments made at the Lake of Geneva it has been ascertained that the rate of progress of sound through water is nearly four times as great as through air. Sound travels through wood or iron still faster — through wood, for instance, it travels from 10 to 16 times as fast as through air, so that it would pass through more than two miles' length of wooden logs in one second of time.

46. Echoes.—Suppose now that I stand in the centre of a large natural amphitheatre, having rocky cliffs all round me, and from this position let me discharge a gun—the noise or impulse will spread from the gun to the rocky cliffs and strike them, but something more will happen after that. The sound when it has struck the cliffs, finding it can get no further, will come back again, and in this particular case it will come back along the very same line that it went, travelling always at the rate of about 1,100 feet per second. The result will be that

a few seconds after the gun has been fired I shall hear the sound that has travelled back from the cliffs just as if another gun had been fired. Now this sound is called an **echo**.

You thus see that in the case of echoes we have the sound or impulse striking an obstacle and then reflected back from it, but it does not always come back in the same direction in which it goes; this depends upon the shape of the surface against which

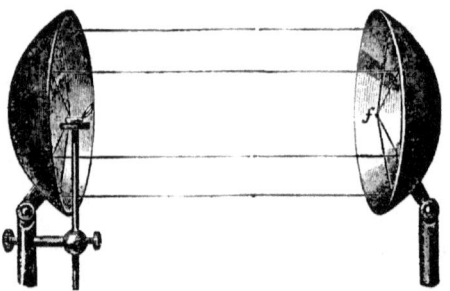

Fig. 22.

it strikes. A very curious experiment is that which is shown in the above figure. Place two large hollow reflectors at some distance from one another, and in a point called the **focus** of the one put a watch, while you place your ear in the focus of the other; you will then hear the ticking of the watch very distinctly, just as if it were close to your ear. The reason of this is that the blows given by the watch to the air strike against the left-hand reflector, and are reflected from it in directions which bring them to the other reflector, from which they are then all reflected into the ear. All this is shown in the figure. This property of sound makes a very nice experiment, but it has sometimes proved inconvenient in practice : for instance, in the Cathedral of Girgenti in Sicily, it is related that the slightest whisper is conveyed from the great western door to the cornice behind the high altar, and that unfortunately the former station was chosen as the place of the confessional. The result was that a listener placed at the other station often heard what was never intended for the public ear, until at length this came to be known, and another site was chosen. The reflexion of sound also explains what takes place in whispering galleries. In that of St. Paul's in London, for instance, a whisper at one side of the dome is conveyed to the opposite side across a very considerable distance.

47. **How to find the number of vibrations in one second corresponding to any note.**— I have told you that when a vibrating body gives the air a small number of blows in one second, we have a deep note, and that when it strikes the air very often in one second, we have a shrill high note : what is called the **pitch** or **tone** of the note depends therefore upon the number of blows which is given to the air in one second. Now we can find out by experiment how many blows in one second correspond to any particular note, and I hope by means of the following figure to make it clear to you how this is done.

You see a large wheel A to the right, which is turned by a handle. Over the circumference or rim of this wheel we have a strong tight strap which passes over the axle of another wheel B. The result is that by means of the strap the axle of the wheel B will go round a great many times for a single turn of A, and the wheel B will itself of course move with

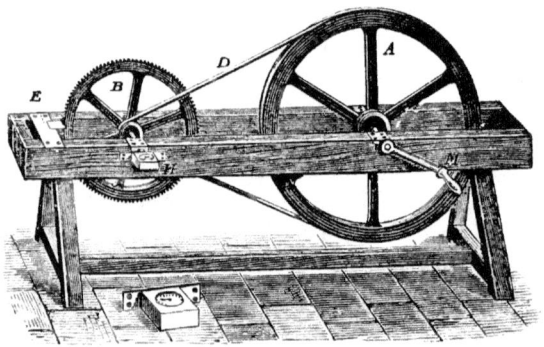

Fig. 23.

its axle—in fact, B may be made to move round very quickly. You see, too, that B is full of small teeth. Now there is a bit of card placed at E against the teeth of B, so that each tooth strikes the card as it passes.

Each time the card is struck we hear a sound, because a blow is given by the card to the air. If there are 100 teeth in the wheel B, there will be 100

blows given to the air in the time that B goes once
round. If B goes round once in a second, 100 blows
will be given to the air, and in consequence 100
sounds will strike our ear in one second, each single
sound of which we shall not be able to distinguish,
but we shall hear an apparently continuous deep note.
Now by driving the handle fast enough I can make B
go round 100 times in a second, and during each
time it will strike the card 100 times; the card will
in this case be struck 100 times 100, or 10,000
times in one second: 10,000 little blows will now
strike the ear each second, and we shall hear a
continuous shrill note.

Now when you wish to find the number of blows in
one second corresponding to a given note, what you
have to do is this. Turn the handle more and more
quickly until the instrument by means of the card
gives you a note precisely of the same pitch as the
note you have got to measure; and when you have
once got the proper speed, keep turning the handle
for some time at the same speed, say for one minute
or more.

Now there is connected with the wheel B a dial
(which is shown separately on a large scale lying
below), and the dial registers how many times the
card has been struck since you began to turn. You
must therefore, when you yourself are turning the
handle steadily at the speed which gives the right
note, get another observer to note the position of the
hand in the dial at the beginning and at the end of

one minute. Suppose he finds out by the dial that
during this minute the card has been struck 60,000
times, this will correspond to 1,000 times in one
second, and hence you conclude that the note given
out is that which corresponds to 1,000 blows given
every second to the air.

HEATED BODIES.

48. **Nature of Heat.**—You have seen that a body
in actual motion may be said to possess energy, and
also that the same may be said of a body in vibration.
You have further seen that a body in vibration does
not, in consequence, move about from one place to
another, but remains at rest as a whole, while,
however, its various particles are moving about alter-
nately forwards and backwards.

You have now to consider bodies in a heated state.
First of all, what is heat? Let us reply by supposing
an iron ball to be put into the fire, and when white-
hot suppose we take it out, put it on the scale-pan of
a balance, counterpoise it, and allow it to cool. Now
if heat be something that has entered into the
ball we should expect that as it cools it will grow
continually lighter. If, however, this experiment be
properly made, it will be found that the iron ball
does not lose weight as it cools, and therefore whatever heat be, its presence has not made the ball one
grain the heavier.

Let me now suppose that I place myself upon a very delicate scale-pan, and while I am there, exactly counterpoised, let some water enter my ear. Of course I shall now be heavier than I was before. Suppose, however, that a sound enters my ear. Will the sound make me heavier? Not one whit. It will strike what is called the drum of my ear, and set it vibrating, and I shall hear the sound, but I shall not be the least whit heavier in consequence of the entrance of the sound into my ear. In fact, while the entrance of water is the entrance of matter, and makes me heavier, the entrance of sound is only the entrance of a kind of vibratory motion, and does not make me heavier. Now may not something of the kind take place in heated bodies? May not the entrance of heat mean the entrance of some kind of vibratory or backward and forward motion, that does not add anything to the weight of the body?

We have strong reasons for thinking that heat is really a kind of vibratory motion, so that when a body is heated each extremely small particle of it is moving about either backwards and forwards or round and round. But these particles are so very small, and their motion so very rapid, that the eye has no means of seeing what really takes place.

Why then, you will say, does not a heated body give out a sound, if, as you tell us, its particles are in a state of rapid motion? Why does not such a body give a series of small blows to the air around

it, just as a body in ordinary vibration does ? We
reply, that a heated body does give a series of blows
to the medium around it ; and although these blows
are such that they do not affect the ear, yet they
affect the eye, and give us the sense of light. You
see now how great a likeness there is between a
sounding body such as a bell and a hot body such as
a white-hot ball. The particles of both bodies are in
a state of rapid motion : those of the bell strike the
air around the bell, and the air conveys the blows to
our ear ; the particles of the hot ball also deal a
succession of blows to the medium around the ball,
and this medium conveys the blows to our eye. Thus
when we experimented on vibrating bodies we used
the ear, but when we experiment on highly heated
bodies we use the eye. And in each case there are
two divisions to the subject : for in vibrating bodies
we have to study in the first place the bodies themselves, how fast they vibrate, in what way they
vibrate, and so on, and in the second place we have
to learn the rate at which the sound they give out is
carried through the air ; so in the case of heated bodies,
we have first of all to study the bodies themselves, and
secondly to learn how fast the rays of light and heat
which they give out travel through the medium.

49. **Expansion of bodies when heated.**—When
a body is heated, it almost always expands ; that
is to say, it gets larger in all directions. To prove
to you that this is the case let us heat a solid, a
liquid, and a gas.

* EXPERIMENT 36.—Let us take (fig. 24) a long
metallic rod held tightly by a screw at one end, B.
The other end is, however, free to expand, and in
doing so it will press against the pointer, P, and in
consequence this pointer will rise ; if, therefore, the
bar expands ever so little, this expansion will be seen
very easily, for it will make the pointer alter its

Fig. 24.

position, and rise up towards the top. Now let us
place two or three lamps beneath the rod and heat it,
and we shall find that the rod expands, and presses
against the pointer so that it rises. If the lamps be
withdrawn, the rod will cool, and in the course of
a few minutes the pointer will have fallen into its
old position.

EXPERIMENT 37.—Here is a hollow glass bulb
which is filled with water ; let us now heat this
glass bulb, and the water will rise in the fine tube
which is attached to the bulb. In this case both the
glass bulb and the water expand, but the water expands more than the glass bulb, and hence it pushes
its way upwards in the fine tube : indeed it expands

with such force that if there were no empty tube into which it might rise it would burst the bulb.

EXPERIMENT 38.—To vary the experiment, let us now take a bladder which is about two-thirds filled with air and heat it over the fire, turning it round so that it may not burn. In a short time the air will have expanded so as to make the bladder appear quite full.

50. **Thermometers.**—You see from all these experiments that the tendency of heat is to make things expand, whether they be solids, liquids, or gases. And now let us particularly consider mercury in a bulb of glass (fig. 25), which like water will become expanded and run up the fine tube when heat is applied. Here we have in reality two things expanding. In the first place the bulb itself expands, so that if you were accurately to gauge the bulb when cold and when hot, you would find it to be slightly larger when hot. The bulb, however, does not expand so much as the mercury, and in consequence the mercury is not content with occupying its old position in the tube attached to the bulb—it must have more room, and to get this it rises in the tube, and, the tube being very fine, a very small expansion of the mercury causes a very considerable rise in the tube, and is thus easily seen by the eye. In fact, the mere warmth of your hand will drive the mercury rapidly up the tube, and a mere breath of cold air will drive it down. An instrument of this kind is therefore very useful for telling whether one thing is hotter or

colder than another, and answers very much better
for this purpose than the feeling of touch. Suppose,
for instance, that we place such an instrument with
its bulb in a vessel of water, and leave it there for a
few minutes, the top of the mercury will then keep a
fixed position in the tube. Let us make a mark, and
note this position accurately. Let us now take the
instrument out of this vessel of water and place it
into another vessel also containing water. If this
water be hotter than the first, the mercury will rise
above the mark which we made—that is to say, the
end of its column will now be higher up; if, however, this water be colder than the first, the mercury
will sink below the mark which we made; and thus
by observing the height of the mercury in the tube,
we can at once tell whether the second vessel of water
be hotter or colder than the first.

An instrument of this kind is called a **thermometer,** and I shall now tell you how a thermometer
is made.

51. **How to make them.**—To make a thermometer, get a glassblower to blow a hollow bulb at the
end of a tube of glass, with a very fine bore, the
other end of this fine tube being open to the air.
Next heat the bulb in a flame; in doing this the air
in the bulb expands (just as it did when we heated
the bladder); but the other end of the fine tube
being open, the expanded air gets out through this
end. Next, before the air has had time to cool,
plunge the open end of the fine tube below the

surface of a vessel containing mercury. The bulb, remember, now contains less air than it did at first, part having been driven out by heat. As this air cools it shrinks into less bulk, and the pressure of the air from without drives up the mercury to occupy the vacant space, just as it drove up the water in the water-pump (Art. 34). Part of this mercury will therefore be driven into the bulb. We have now got a little mercury in the bulb, and we next take the bulb with the mercury in it and heat it well above the flame of a lamp—bulb, tube, and all. The mercury will soon begin to boil, and its vapour will drive out the air before it, until bulb and tube will both be filled with the vapour of mercury. When this is done, we plunge the open end of the tube once more into a vessel of mercury. As there is now no air in the tube or in the bulb, but only vapour of mercury, when this cools it will condense and there will be a vacuum, and the mercury in which the instrument is plunged will be driven up by the pressure of the outside air until it fills both tube and bulb. We have thus filled both tube and bulb with mercury, and now before it has cooled we seal the open end by melting the glass, so as to keep the air out, and this part of the process is then complete.

Having thus got our thermometer tube, we plunge it when sufficiently cool into a box containing pounded ice which is in the act of melting. The column of mercury of course falls in the tube because the ice is very cold : (you have been told that the column of

mercury falls when the bulb is plunged in a cold substance). When the mercury ceases to fall, mark off
with a file the position of the top of its column in the
tube: this is the position which the top of the column
will always occupy when the instrument is put into
melting ice, or something equally cold.
Having done this, next take the thermometer tube and plunge both bulb and
tube into boiling water, and mark off the
position of the top of the column as before. The column will now, of course, be
very high, for the mercury will have
expanded very much in consequence of
the hot water. You have now got two
marks on your fine tube—the one denoting the position of the top of the column
of mercury when you plunge the bulb
into melting ice, the other the position of
the top of the column when you plunge
the bulb and tube into boiling water.
You will afterwards learn that the heat
of boiling water is not quite constant, but
for the present we may regard boiling
water as having a fixed heat.

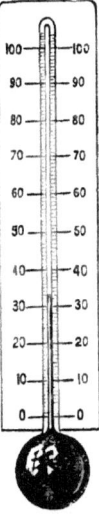

Fig. 25.

Having thus got two points marked or scratched
with a file upon the tube of the thermometer corresponding to the freezing and to the boiling points of
water, the next operation is to divide the whole
distance between these two points into 100 equal
parts. This is done by coating the whole tube with

wax, and then making scratches in the wax-coating with the point of a needle at the proper places. If we then dip the whole tube into a solution of hydrofluoric acid, this will not affect the wax, but it will affect the glass where the point of the needle has cut through the wax. After the tube has been taken out of the solution we shall therefore find that all the lines which we made with the point of the needle have eaten into the glass by help of the acid, and form, in fact, a scale of lines by the aid of which we may rise from the freezing to the boiling point of water through 100 steps or stairs, each step denoting something hotter than the one below it, and colder than the one above it.

Finally, let us call the lower step 0 degree, the upper step 100 degrees, and let us also number every ten steps between these two, and our thermometer is complete.

Such an instrument is called a **centigrade thermometer**, which means a thermometer with **a hundred steps** ; and as this is the most convenient form of graduation, we shall always use it.

If a substance be of such a heat that when a thermometer is placed in it the end of the column rises to 10 or 20 or 30, we say the **temperature** of the substance is 10 or 20 or 30 degrees, and so on. Melting ice therefore has the temperature of 0 degree (written 0°) on the Centigrade scale, and boiling water the temperature of 100 degrees (written 100°) on the same scale ; 20° is a good summer heat, and 35° is about the heat of our blood, or blood-heat. In fact, such an instrument gives us a very accurate means of measuring temperature.

52. **Expansion of Solids.**—By a method similar to that of Experiment 36, only more accurate, we have found out how much rods made of glass or of metal expand between the freezing and the boiling points of water, that is to say between 0° and 100° of the thermometer, and the results are exhibited in the following table :—

				Expansion between the freezing and the boiling points of water of a rod 100,000 inches long.	
Glass	.	.	.	85	inches.
Copper	.	.	.	171	,,
Brass	.	.	.	188	,,
Soft iron	.	.	.	120	,,
Cast iron	.	.	.	109	,,
Steel	.	.	.	114	,,
Lead	.	.	.	282	,,
Tin	.	.	.	196	,,
Silver	.	.	.	192	,,
Gold	.	.	.	144	,,
Platinum	.	.	.	87	,,
Zinc	.	.	.	298	,,

53. **Expansion of Liquids.**—Liquids expand more than solids when you increase their heat, but you cannot make an experiment upon a liquid rod, because a liquid cannot form a rod. In this case let us take a definite measure, such as a pint, and find

what would be the overflow in pints of a liquid that occupied 100,000 pints at the freezing-point of water if it were raised to the boiling-point.

Now, if 100,000 pints of mercury were heated from 0° to 100°, or from the freezing to the boiling point, there would be an overflow of 1,815 pints; and if 100,000 pints of water were heated between the same range, there would be an overflow of 4,315 pints.

It is found by such experiments that

Liquids expand more than solids for the same increase of temperature, and that liquids expand more rapidly at a high than at a low temperature.

54. **Expansion of Gases.**—Gases expand through heat, and that to a great extent; but here we must bear in mind that other things besides heat will make gases expand. You remember the india-rubber ball that was put into a receiver and began to expand when the air was withdrawn from the receiver (Experiment 25). When, therefore, we wish to see how much a quantity of gas expands through heat, we must take care that the air which surrounds the gas does not change its pressure : in other words, we must take a bladder with some air in it, and find how much it expands when heated in the open air—that is to say, under the constant pressure of the atmosphere—between the freezing and the boiling points of water.

When this is done, it is found that if a bladder not completely filled with air have a volume equal to 1,000 cubic inches at the freezing-point, its volume at the boiling-point will be 1,367 cubic inches.   If therefore we have a large quantity of ice-cold water in a vessel and force this bladder containing 1,000 cubic inches beneath the water, we shall find the water rise in the vessel through a space denoting 1,000 cubic inches—this being the increase of volume due to the bladder.   But if we take the same vessel, only filled with boiling water, and plunge the bladder into it, we shall find the water rise through a space denoting 1,367 cubic inches—this being the volume of the bladder at this temperature.

55.   **Remarks on Expansion.**—Liquids and solids expand with immense force.   If you were to fill an iron ball quite full of water, shut it tightly down by means of a screw, and then heat the ball; the force of the expansion would be great enough to burst the ball.

In large iron and tubular bridges allowance must be made so that the iron has room to expand; for in the middle of summer the bridge will be somewhat longer than in the middle of winter, and if it has not room to lengthen out, it will be injured by the force tending to expand it.   There is an arrangement for this purpose in the Menai Tubular Bridge.

We take advantage of the force of expansion and contraction in many ways—for instance, in making carriage wheels.   The iron tire is first made red-hot, and in this state is fitted on loosely upon the wheel; it is then rapidly cooled, and in so doing it contracts, grasps the wheel firmly, and becomes quite tight.

56. **Specific Heat.**—Some bodies require a greater amount of heat than others in order to raise their temperature one degree. The quantity of heat required to raise a pound weight of any substance one degree is called its **specific heat.** Water has a very great specific heat; that is to say, it requires more heat to raise a pound of water one degree than it does to raise almost any other substance. The heat required to raise a pound of water one degree will raise through one degree 9 lbs. of iron, 11 lbs. of zinc, and no less than 30 lbs. of mercury or gold.

EXPERIMENT 39.—To convince you of the great specific heat of water, let us take 2 lbs. weight of mercury and heat it to 100°, or the boiling-point of water, and let us then mix it with 1 lb. of water at an ordinary temperature. Now note the height of a thermometer placed in the water both before and after the mixture, and you will find that it has hardly risen more than 5° in consequence of the hot mercury being poured in.

57. **Change of state.**—You have already heard about the three states of matter—the solid, the liquid, and the gaseous. I have now to tell you that substances when heated pass first from the solid to the liquid, and then from the liquid to the gaseous state. You are told in the Introductory Primer that ice, water, and steam have precisely the same composition, and that ice becomes water if it be heated, while water becomes steam if we continue the heat. The very same change will happen to other substances if

we treat them in the same way. Let us, for instance, take a piece of the metal called zinc and heat it ; after some time it will melt, and if we still continue to heat it, it will at last pass away in the shape of zinc vapour. Even hard, solid iron or steel may be made to melt, and to be driven away in the shape of vapour ; and by means of an agent called electricity (of which more hereafter) we can probably heat any substance suffi· ciently to drive it away in the state of vapour or gas.

We cannot, however, cool all bodies sufficiently to bring them into the solid or even into the liquid state. Thus, for instance, pure alcohol has never been cooled into a solid ; but we know very well that all we have to do is to obtain greater cold in order to succeed in freezing alcohol. In like manner, we have never until quite recently been able to cool the atmospheric air sufficiently to bring it into the liquid form ; but here, as before, the thing required is great cold. You must not, however, imagine from what I have said that cold means anything else than the absence of heat. A cold body is a body which has little heat, and a still colder body has still less heat ; but even the coldest body which we can produce has a little heat left. Do not be guided in this respect by your feeling of touch. Two bodies may be of the same temperature, as shown by the thermometer ; and yet the one may feel much colder to you than the other ; and if you keep one hand for some time in very cold and the other in very hot water, and then plunge them both into water of ordinary heat, this water will seem hot to the one hand and cold to the other. Do not therefore be guided by anything else than the thermometer, or imagine that cold is anything else than the absence of heat.

To return to our subject. Probably all bodies, if we could cool them enough—that is to say, take away enough of their heat—would assume the solid state; and then, when each was again heated sufficiently, it would become liquid, until at last, if still heated, it would be driven off in the shape of gas or vapour. There would, however, be a great difference between the different bodies in the ease with which they would yield. Ice soon melts if we apply heat; tin or lead require to be heated to 200 or 300 degrees before they will melt; iron is more difficult to melt than lead; and platinum is more difficult than iron. A body very difficult to melt is called **refractory**.

In the following table we have the temperature at which some of the most useful substances begin to melt.

Ice melts at .	0°
Phosphorus .	44°
Spermaceti .	49°
Potassium .	58°
Sodium	97°
Tin	235°
Lead .	325°
Silver .	1,000°
Gold .	1,250°
Iron .	1,500°

Platinum is so difficult to melt that we cannot tell at what temperature it does so. And carbon is still more difficult to melt—indeed in the very hottest fire the coal or carbon is always solid; and no one ever heard of the coal melting down and trickling out through the furnace bars.

We thus see that the same sort of cnange takes place in all bodies through heat; that is to say, if we could reach a temperature sufficiently low, all bodies would become solid like ice, and if we could reach one sufficiently high, all would become gaseous like steam: in fact, the change that takes place is always of the same kind, and we cannot do better than use water as a type of all other things in this respect, and study the behaviour of this substance under heat, beginning with its solid state when it appears in the shape of ice.

58. **Latent heat of Water.**—Let us take some very cold ice, pound it into small pieces, and put the bulb of our thermometer into this pounded ice. Let us suppose that the reading of our instrument shows a temperature 20 degrees below the point we call 0°. Now let us heat the ice, and its temperature will rise like that of any other solid under like circumstances until it comes to 0°, but at this point it will stop, and rise no further as long as any ice remains. What then does the heat do if it does not raise the temperature above this point? We reply, it **melts the ice.** At first the heat is wholly spent in raising the temperature of the very cold ice, but when this

temperature has reached 0° the heat has quite a different office to perform ; its power is now wholly spent in melting the ice, and when the ice is all melted, the water has only the temperature 0°, being no hotter than melting ice. In fact, water at 0° is equal to ice at 0°, together with a large amount of heat, which we call **latent heat** because it does not affect the thermometer.

EXPERIMENT 40.—You may prove this by putting some pounded ice into a tin pan and heating it over a lamp until there is only a little ice left. If you then plunge a thermometer into the melted ice, you will find that the temperature will hardly be above 0°, or in fact the melted ice will be as cold as the ice before it was melted.

59. **Latent heat of Steam.**—We have now changed our ice into water, and if we continue to heat this water its temperature will rise in the ordinary manner, like that of other bodies, until it reaches the boiling-point or 100°. Its temperature will then stop rising, and if we continue to heat the water we shall only convert it into steam of which the temperature is 100° and no more. In fact, just as it took a large amount of heat to convert ice at the freezing-point into water at the freezing-point, so does it take a large amount of heat to convert water at the boiling-point into steam at the boiling-point. So that we are entitled to say— steam at 100° is equal to water at 100°, together with a large amount of heat which we call latent, because it does not affect the thermometer.

EXPERIMENT 41.—You may prove this by boiling some water in a flask and putting the thermometer first into the boiling water and then into the steam. They will both be found to have the same temperature, or, in other words, steam is no hotter than boiling water.

Thus you see that ice requires latent heat to bring it into water, while water again requires latent heat to bring it into steam. Now we can measure how much heat it will take to bring a pound of ice at 0° to a pound of water at the same temperature, and we find that it will take as much heat to do this as it would to raise 79 pounds of water one degree in temperature, and this is what we mean when we say that the latent heat of water is equal to 79. In a similar manner it has been found that the latent heat of steam is 537; that is to say, it will take as much heat to change a pound of water at 100° into steam of the same temperature as it would to raise 537 pounds of water one degree in temperature.

It thus takes a good deal of heat to melt ice, and it therefore takes a good deal of time to do so. Indeed it is much better that this is the case, for what would happen if ice at the melting-point were to change into water at once when heated ever so little? It would render uninhabitable a large part of the globe, for the ice of the mountains would on some fine spring day be at once liquefied, and the water would rush down in such overwhelming torrents as to sweep everything before it, and large tracts of

low-lying land would be buried under water.   In like
manner, it is much better for us that it takes a
large amount of heat to convert water at the boiling-
point into steam ; for suppose that water at this point
were at once converted into steam by heating it ever
so little, there would then be an explosion in every
tea-kettle and in every boiler, while a steam-engine
would be an utter impossibility.

You have already been told that steam is a gas like
air, and you have learned in the Introductory Primer
that you cannot see true steam.   When a kettle is
boiling rapidly, you may have noticed that you do
not see anything quite close to the spout of the kettle,
but about half an inch beyond it you see a cloud.
Or, again, when a locomotive gives out its steam you
do not see anything quite close to the funnel, but
a little distance above it you see a cloud.   Now this
invisible thing that comes out is true steam, but the
visible cloud consists only of very small drops of
water, formed from the steam as it cools ; it is not
therefore steam, but water.   True steam is invisible,
like the air or any other gas.

60. Ebullition and Evaporation.—I have now
told you something about the steam which is given
out when water boils.   I do not, however, mean to
say that no steam is given out before it boils, for this
would be contrary to fact : all of you must have
noticed that a pan of water put on the fire gives out
steam long before it begins to boil.   Doubtless, too,
you must have noticed that any wet thing or thing

full of water gets dry near the fire—that is to say, its water goes away in the shape of steam. Now when steam or vapour (for both words mean the same thing) is given out by water which is not boiling we call it **evaporation**, but if the water boils we call it **ebullition**. The difference is simply this. When you heat water over the fire, the heat has at first two things to do. It has in the first place to heat the water, and in the next place it evaporates part of the water ; but when the temperature of the water has risen to 100° or the boiling-point, the water cannot be heated above this : all the strength of the fire is then spent in converting the water into steam, and this steam escapes not only from the top of the water but from the very bottom also, so that we hear a noise which we call boiling as the bubbles of steam rise through the water and escape into the air.

61. **The boiling-point depends on pressure.—** I have now to tell you that the temperature or heat at which water boils is not a perfectly fixed point like that of melting ice, but depends upon the pressure of the air. If the pressure of the air be lessened, water will boil below 100°. You remember you were told that the pressure of the air is less at the top of a lofty mountain than at the bottom, because at the top you have a less depth, and therefore a less weight or pressure, of air above you. Well, at the top of Mont Blanc in Switzerland, which is three miles high, water will boil at 85° ; and if a traveller were to try to boil an egg in a pan at the top of Mont

Blanc, he might boil it for hours, but it would not harden, because 85° is not high enough to harden the white of an egg.

On the other hand, if we were to boil water at the foot of a very deep mine the boiling-point would be considerably above 100°.

EXPERIMENT 42.—You will see by the following very simple experiment that the temperature of the

Fig. 26.

boiling-point depends upon the pressure of gas or air upon the surface of the water. Let us take a glass flask and fill it half full of water, then cause the water to boil for some time, until the steam has driven out all the air from the upper part of the flask, so that we have only water, and the vapour of water in the flask. Now cork it tightly, and withdrawing it from

the lamp, invert it as in fig. 26. When it has
ceased boiling, take a sponge and pour some cold
water on the flask, when boiling will again begin.
The reason of this is, that before the cold water was
poured on there was a considerable pressure of
vapour upon the water of the flask, and this pressure
kept it from boiling, but the effect of the cold water
was to condense this vapour, and therefore to lessen its
pressure; and since water boils more easily at a low
pressure than at a high, the water in the flask began,
as you saw, immediately to boil.

Before leaving this part of our subject, I ought to
tell you that some bodies expand while others contract in the act of melting; that is to say, in passing
from the solid to the liquid state.

EXPERIMENT 43.—Here is some ice, for instance,
which is lighter than water, as you will see by the
fact that the ice is at present floating upon the water.
In passing from ice to water there is, therefore, a
great contraction of substance, and in passing from
water to ice—in the process of freezing—there is a
great expansion. This expansion takes place with
great force, and if a thick iron vessel be filled with
water and then shut by a stop-cock, by causing the
water to freeze you may burst the iron vessel. Steel
and probably cast iron contract like ice when they
melt, or, which is the same thing, expand like water
when they freeze or get solid. Thus, for instance, a
piece of white hot scrap steel will float in a bath of
melted steel, and it is believed that a piece of red-hot

D

cast iron will float in liquid cast iron. On the other hand, gold, silver, and copper expand when they melt, and contract when they become solid ; they will not, therefore, run into the crevices of a mould, and therefore coins made of these metals cannot be cast in a mould, but must be stamped.

All substances, however, expand very greatly when they are converted into gas, and a cubic inch of boiling water will be converted into steam occupying nearly 1,700 cubic inches.

62. **Other effects in Heat.**—You have now seen that heat expands bodies or makes them larger, and that it also causes them to change their state, passing from solids to liquids and from liquids to gases as the heat continues to be applied. You have seen how powerful an agent heat is; how the strongest and hardest bar of iron will by it be changed into a white hot mass as soft as treacle, and if heated still more will be driven off in the shape of gas.

Heat affects bodies in many other ways, and more especially it promotes the operation of chemical attraction. Thus at a low temperature coal will not combine with the oxygen of the air, and we may keep our coals for any length of time in our coal-cellar. But when heat is applied combination takes place ; and as this combination in its turn produces heat, the process of combination goes on, and the coal is said to burn.

In like manner in the experiment (Chemistry Primer, Art. 6) where sulphur and copper combine together, heat is first of all applied in order to promote combination, but when this has begun heat is generated, and the process goes on of itself, without requiring any more heat from a lamp.

63. **Freezing Mixtures.**—Again, chemical union, you have been told (Chemistry Primer, Art. 7), produces heat, and this is always true; nevertheless sometimes two substances which have a tendency to form a solution mix together with the production of cold and not of heat. Thus common salt and snow have a tendency to form a solution, and they will do so with the production of very considerable cold, or, to speak more correctly, with the absorption of a very considerable quantity of heat.

EXPERIMENT 44.—To prove this, let us rapidly mix some melting ice or snow and some salt together, and place in the mixture the bulb of our thermometer. The mercury in the tube will soon fall below 0°, thereby showing that this mixture is colder than melting ice.

Now what is the reason of this? it is to be found in the fact that after these two substances have become mixed together we have a liquid and not a solid—in fact, we have strong brine. Now you have been told that heat is swallowed up, or becomes latent, when bodies pass from the solid into the liquid state—for instance, when ice becomes water. The brine, therefore, being a liquid, swallows up part of the heat of the snow and salt, and the consequence is that we have a very cold liquid as the result of the union of two solid bodies. Thus when two solid bodies dissolve each other, we have very frequently a lowering of temperature on account of the heat which is swallowed up by the liquid. Such bodies are said to form **freezing mixtures**.

In like manner, if we have a liquid that evaporates very fast we find it to be intensely cold, because in order to become a vapour or gas it requires a great deal of heat, and gets it where it can: thus if you drop some ether upon your hand it feels very cold, and soon flies away in the shape of gas; in fact, it has robbed your hand of a large quantity of heat in order to produce this vapour or gas. Very low temperatures, very intense cold, may sometimes be produced by causing certain liquids to evaporate very rapidly.

EXPERIMENT 45.—To prove this let me pour some water into a shallow vessel, and place it along with a pan containing strong sulphuric acid under the receiver of the air-pump, and exhaust the air. As the pressure of the air is withdrawn the water will evaporate very rapidly, and in order to do so will take away so much heat from its own substance that it will be turned into ice.

64. **Distribution of Heat.**—Let us now proceed to another part of our subject, and consider the tendency which heat has to distribute itself.

A hot body will not always remain hot, but it will part with its heat to the colder bodies that are around it; and it will always insist upon doing this, but it will do it in different ways according to circumstances.

EXPERIMENT 46.—For instance, let us put a poker into the fire; some of the heat of the fire gets into that part of the poker which is in the fire, and this passes along the poker until it heats that end which is farthest away from the fire, and you will at last find it too hot to touch. This passage of heat along the poker is called **conduction of heat.**

EXPERIMENT 47.—Again, let us take a flask two-thirds full of water, and heat it from below. As the lower particles of water are heated they expand, and therefore get lighter; they consequently rise to the top for the same reason that a cork rises in water, and are replaced by colder and, therefore, heavier particles from above. A new set of particles are thus continually subjected to the heat of the lamp, and in process of time the whole water will get heated and begin to boil. This process is called **convection of heat.**

Neither of these processes will, however, account for the heat that reaches us from the sun. Whether in conduction or convection the heat is carried by means of particles of solid or liquid matter, but we have reason to think that there are no such particles between us and the sun, while we know that the sun's light and heat takes less than eight minutes to come from the sun to us over a distance of 90 millions of miles. Evidently, then, the heat which comes to us from the sun moves with an immense velocity, and does not reach us in virtue of heating up the particles between the sun and ourselves. In fact, in a very cold day, when the air is very cold and anything but heated, the sun's rays may be very powerful. Now the process by which heat comes to us from the sun or any other very hot body is called **radiation of heat.**

We have thus three very different ways in which a heated body communicates its heat to a cold one; namely, conduction, convection, and radiation. Let us now consider these in order.

65. **Conduction of Heat.**—We have spoken about thrusting a poker into the fire, and told you that at last the other end of the poker will be too hot to hold. But if, instead of a metal poker or rod, a glass or stoneware rod were thrust into the fire, the other end of this rod would never get very hot, because stoneware does not conduct heat nearly so well as metal.

Wool and feathers are still worse conductors, and this is why these substances have been provided by nature as the clothing of animals; for the heat of an animal is generally greater than that of the surrounding substances, and this heat is not readily conducted off through the garment of wool, feathers, or fur, with which the animal is clad. So in the case of boilers of engines; when we wish to keep in the heat, we supply them with steam jackets or coverings made of a non-conducting substance.

A bad conductor may be used not only to keep in heat, but also to keep it **out**; flannel, for instance, may be used to wrap round our bodies in order to

keep the heat in, or it may be used to wrap round a
block of ice which we wish to preserve in order to
keep the heat out. In fact, heat cannot readily pass
through flannel whether it be going from within
outwards or from without inwards.

EXPERIMENT 48.—It is very easy to show you that
different substances have different conducting powers
for heat. You see, as in the figure, two rods or
wires, one of copper and one of iron, with their ends
together, at which they are heated by means of a

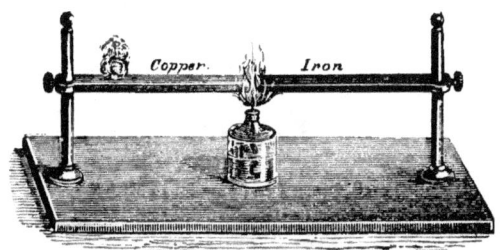

Fig. 27.

lamp. After the lamp has burned for some time, let
us take two little bits of phosphorus, and place one of
them at the end of the copper rod furthest away from
the flame. It will soon take fire. Now place the
other piece on the iron rod at the same distance from
the lamp as the burning phosphorus, and it will not
take fire. This shows us that the heat of the lamp
is conducted more powerfully along the copper than
along the iron.

The conduction of heat explains the action of the safety lamp which was devised by Sir Humphry Davy for the use of miners, but this very useful lamp has already been fully described in the Chemical Primer (Art. 41).

66. **Convection of Heat.**—If we take a vessel full of water, and float on its surface a vessel full of

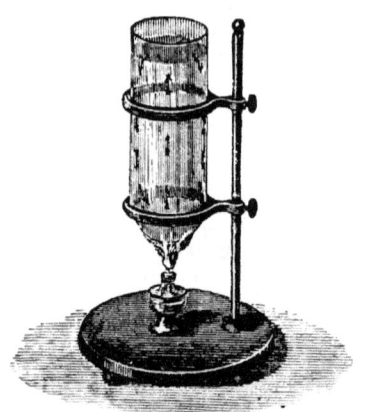

Fig. 28.

boiling oil, we shall find that the heat of the oil will be conducted very slowly indeed downwards through the liquid; in fact, a few inches down, the rise of temperature will be hardly perceptible. But if instead of heating the vessel with water in it from above we heat it from below, as in the figure, we shall find that in a very short time the whole water will be heated and begin to boil. In fact, as we have already stated, the heated particles getting lighter rise, and are replaced by colder and heavier particles from above, so that we have a current as is shown by the arrows in the figure, the heated water ascending in the middle and the cold water coming down the sides.

We have several good examples of convection in nature; for instance, in a lake which is cooled at its surface by the action of intense cold. The surface particles are first cooled, and getting heavier they sink down and are replaced by lighter and warmer particles from beneath, so that in a short time the whole body of water becomes cooled down to a temperature about $4°$ above the freezing-point; after that temperature, the water, contrary to the usual practice of things, expands when further cooled instead of contracting; and when ice is formed, this ice, being decidedly lighter than water, floats on the top.

Now, had ice been heavier than water, it would have fallen down to the bottom as it was formed, a fresh surface would thus have been exposed, and the whole lake would soon have become one mass of ice. But as it is, the cold can only freeze the second layer of water through the ice of the first, and this is a very slow process, so that there is no danger of a lake being permanently frozen.

In the air again we have strong convection currents due to heating; for it is on this account that the hot air of a fire goes up the chimney, being replaced by the cold air from the room ; and we have the very same thing on a large scale in the great system of winds, for at that part of the earth called the equator, where the sun is most powerful, the air when heated mounts up just as the air of a fire mounts up the chimney. This air is then replaced by currents blowing along the surface of the earth from the poles or colder portions of the earth. We have thus at the equator, a system of upward currents which carry off the hot air to the poles in the upper regions of the air, and we have also currents blowing along the surface of the earth, which bring back this air when cooled to the equator. These surface-currents blowing from the poles to the equator are called the **trade winds.**

67. **Radiant Heat and Light.**—The third method by which a hot body parts with its heat is by radiation, and it is in virtue of this process that the heat of the sun reaches our earth. We need not, however, go farther than our own firesides to get an example of the process. If we stand opposite a strong fire, we find our faces and our eyes suffering from the heat. Even a kettle containing hot water gives out radiant heat, although the rays of heat from it cannot pierce the eye and impress it with the sense of light like those from the fire or from the sun. Thus when you heat a body such as a ball of clay, something of the following kind takes place. The body begins at once to rise in temperature, and in consequence to give out rays of heat, but those

rays are dark rays, and do not affect the eye. As
the heating process goes on, a few of the rays which
it gives out begin to affect the eye, and the body
becomes red hot; it next acquires a yellow heat, next
a white heat, and last of all it glows with an intense
light resembling the sun. Let us now devote ourselves for a short time to the study of these bright
rays which a hot body gives out.

68. **Velocity of Light.**—Römer, a Danish astro-
nomer, was the first to find out the velocity with
which light travels through space. To understand
what this means let us remember what takes place
when a distant gun is fired off. We see a flash, and
then some seconds after we hear a report. Evidently
then the sound does not reach the ears at the very
moment when the gun is fired, because it lags behind
the light. But does the light reach us at the very
moment? may not both light and sound start from
the cannon at the same moment, and each take some
time to get to us, the light winning the race and
coming in first? This point can only be decided
by observation and experiment, and it was by observation that Römer found it out. There is a large
planet called Jupiter, which is sometimes very far
from us and sometimes comparatively near, and this
large planet has several satellites, or small attendants,
one of which passes across the disc or surface of
Jupiter at regular intervals, so that when we use a
powerful telescope we can see the small satellite like a
black body crossing the large disc of the planet.

Now Römer found that when Jupiter was very far
away from us the satellite seemed to be later in
crossing than it ought, and he argued from this that
we on the earth do not see the crossing of the satellite
over the disc or surface of Jupiter at the very
moment when it takes place, but that light takes
some time to get from Jupiter to our eyes, just as
the report of a distant gun takes some time after
the explosion to reach our ears.

You thus see that light as well as sound takes time
to travel, only light travels much faster than sound
—light travels at the enormous rate of 186,000 miles
a second, while sound creeps along at the rate of
1,100 feet in the same time. Light only takes 8
minutes to come from the sun to us, although the
distance is 90 millions of miles. If, therefore, the
sun were to be suddenly extinguished, we should not
find it out until 8 minutes afterwards.

Do not, however, suppose that light consists of
small particles shot out by hot bodies, and flying
through space at the enormous rate of 186,000 miles
a second.   If this were the case, we should be
knocked to pieces by a ray of light. A ray of light
may be said to enter the eye just as sound may be
said to enter the ear. We have already explained
that when we hear the report of a gun, it does not
mean that small particles of air travel all the way
from the gun to our ear. And so when we view a
ray of light it does not mean that a small particle is
shot from the bright body into our eye. An impulse

or wave in each case passes over the medium between us and the body, and the blow goes from particle to particle after the manner we have already explained in the experiment with ivory balls (Art. 44).

69. **Reflexion of Light.**—When light strikes a polished surface of metal, it is reflected from it. If you hold a lighted candle before a mirror, you will see the image of the candle in the mirror, which

Fig. 29.

means that the rays from the candle strike the mirror and are reflected from it to your eye, just as if they came from the mirror itself and not from the candle.

EXPERIMENT 49.—In order to understand how reflexion acts, let us take a horizontal polished metallic surface—that is to say, let us pour mercury into a shallow flat-bottomed vessel. Now place a bent tube open at the bottom above the mercury as in fig. 29, and let the light of a candle enter the tube at the right end : if we place our eye at the left end, we shall see the light from the candle as it comes reflected from the surface of mercury.

In this experiment, therefore, the light of the candle goes down the one tube, strikes the surface of

mercury and then ascends the other tube to the eye.
But in order that the light may do this, two things
are necessary. In the first place, the two tubes must
have the same inclination or slope ; and secondly,
the one tube must be exactly opposite the
other, so that if they were suddenly to fall flat down
they would be in a line with one another. Whenever,
therefore, a ray of light strikes a polished surface,
the reflected ray rises from the surface with the same
slope as the ray that strikes the surface falls towards it,
and both rays, if you could imagine them squeezed flat
against the surface, would be found to make one line.

You cannot understand the laws of reflexion
completely without geometry, but the following figure
will perhaps enable you to do so to some extent. In
the figure, A is supposed to be a bright point giving
out light, and M M is a mirror. Let A B, A B' be two
of the rays of light from A, striking the mirror at B
and B'. These will then rise into the eye of the
observer in the directions B D, B'D', the falling slope of
the ray A B being equal to the rising slope of B D, and
the falling slope of A B' equal to the rising slope of
B'D'. Now if you imagine the direction of the two
rays B D, B'D', prolonged beneath the mirror, they
would meet at A', a point as much below the mirror,
as the bright point A is above it. To the eye, therefore,
the rays will in point of fact appear to proceed from
A', so that the apparent position of the reflected
image A' is as much behind the mirror as the bright
point A is itself before it.

Whenever, therefore, you stand in front of a
mirror, you see your own image in the mirror as
much behind the mirror on the other side as you
yourself are in front of it; if you go close to the
mirror, the reflected figure goes close also, if you
draw back the reflected figure draws back, and so on.
You will, however, notice the difference—namely,
that your right hand is the left hand of the
image, and your right side generally the left

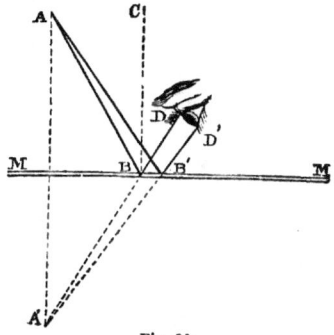

Fig. 30.

side of the image, but in other respects the image
is precisely a copy of yourself.

In fig. 31 you see in the lower part the image of
the upper part, and you notice how, in the image,
the letters go from right to left, and not from left to
right,

When the bright reflecting surface is not flat,
curious images are sometimes produced. Take, for
instance, the bright surface of mercury in the bulb of

the thermometer and look into it. You will there see a very small distorted image of yourself, and indeed of the whole room, only the far away parts of the room will be exceedingly small.

Take again a couple of bright concave mirrors like those of fig. 22, only, instead of putting a watch at the focus of the one mirror, and your ear at that of the other, place a red-hot ball in the one focus, and your

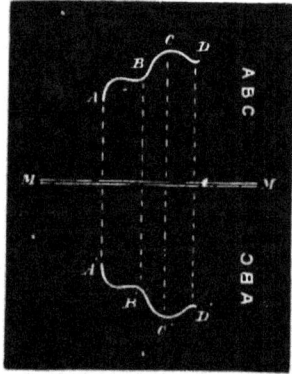

Fig. 31.

hand in the other focus, and you will soon find it too hot. Indeed, if you had two large reflectors of this kind and had a fire burning in the focus of the one, you might cook a beef-steak in that of the other, even though the two reflectors were fifty feet apart. The reason is that the rays of heat from the fire in the one focus strike the mirror near it, and are reflected from it in lines that bring them to the other reflector, and they are then again reflected in such lines as to bring them all together into the focus of this reflector. We thus have, as it were, the fire itself burning at the one focus, and an image of the fire at the other, the image being sufficiently hot to cook a beef-steak.

70. **Bending or refraction of Light.** EXPERI-
MENT 50.—Put a small, heavy body at the bottom of a stoneware or pewter jug, and put your eye in such a position that the side of the jug just hides the body

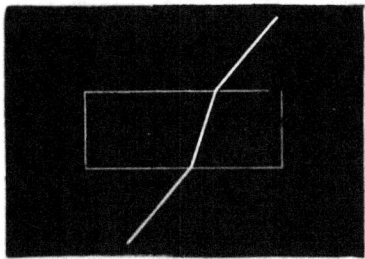

Fig. 32.

from your eye; then let some one fill the jug full of water, and the small body at the bottom will now become visible. Why is this? It is because the ray of light from the small object at the bottom of the water after it leaves the surface of the water is bent in a different direction, so that you can in fact see it round a corner; and if the small body at the bottom were a little fish, it could also see you.

It thus appears that if a slanting ray of light
strikes a surface of water, it is bent so as to be less
slanting after it enters the water ; or again, if a ray
of light comes out from the water, it is bent so as to
be more slanting after it enters the air. The same
thing would happen if the ray of light entered a
surface of transparent glass instead of a surface of
water,—a slanting ray would become less slanting
after it entered the glass. If you had a flat, thick
piece of glass, the ray of light would take the course
that is shown in the preceding figure, in which we see
that its path before it enters the glass, and its path
after it leaves the glass, are in the same **direction**
(though not in the same **line**), while, however, its
path in the glass is quite different.

Suppose, however, that the piece of glass is not flat
but shaped like a wedge ; in fact, that it stands

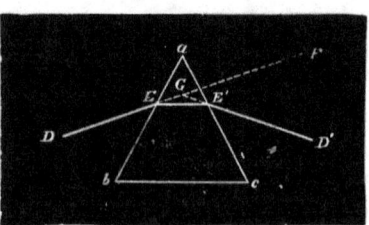

Fig. 33.

straight up above the page on a bottom like fig. 33,
and that when viewed standing up it has the appearance of fig. 34. Such a piece of glass is called a

prism. Let us now see in what manner a ray of
light will be bent in passing through
a prism. This is exhibited in figure
33, from which you see that the ray
is bent towards the thick part of
the prism ; in fact, the direction of
the ray is entirely changed.

You thus see that whenever a ray
of light passes through a wedge-

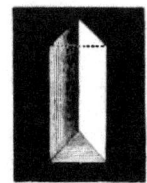

Fig. 34.

shaped piece of glass, it is bent towards the thick
part of the wedge.

71. **Lenses, images given by them.**—Now let
us vary the shape of the piece of glass in the following
manner. Let the piece of glass be circular like a cake,
only thickest in the middle and thinnest all round the
edge ; in fact, appearing like a circle if viewed in one
direction, but if viewed endwise appearing like the
following figure.

Such a piece of glass is called a **lens**. Now let a
bundle of rays of light from a distance fall
upon the lens. What will happen ? The
lens will act like a circular wedge ; it is
in truth a circular wedge, and being thickest
in the middle the rays of light will be bent
towards the middle all round the lens. In
fact, the rays of light will come to a point,
or nearly so, as will be seen from the following figure.

Now suppose that when the sun is shining, you
place a lens so as to allow the rays from the sun to
strike it full on the surface, these rays will be brought

Fig. 35.

to a point, or nearly so, on the other side of the lens;
and if you place a sheet of paper at this point, you
will see a small bright image of the sun, which will
be so intensely hot as to set fire to the sheet of paper;
in fact, the lens will now act as a burning-glass.

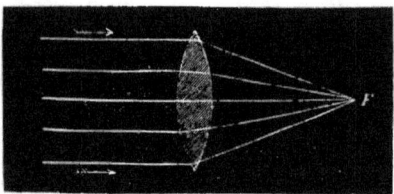

Fig. 36.

EXPERIMENT 51.—Such a lens will give an image of
anything else as well as of the sun ; for instance, I
have here an arrangement by which the rays of light
from a candle are allowed to fall full upon a lens, and
I obtain upon a piece of oiled paper placed on the other
side of the lens an image of the candle, only as you
see upside down.   In fact, if you place anything at all
bright in front of a lens some distance off, behind the
lens you will get a small image of this thing.   If you
place your face in front of the lens, behind the lens
there will be a small image of your face.   Now this
is precisely what the photographer does.   He has a
black box with a lens at one end of it, such as you see
in the following figure.   He points the lens to a
landscape or to the face of a person, and in the dark
box there is a little image of the landscape or of the

face, which he first of all allows to fall upon ground
glass, so that he can see it and know if it be right.
He then takes out this ground-glass plate and puts in
its stead a plate of glass having its surface covered
over with a peculiar substance that is acted on by
light. The image inside the box now falls right upon
this sensitive chemical substance, and the bright
parts of the image act upon and change the nature of
the surface, while, however, the dark parts do not

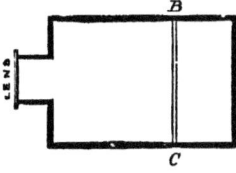

Fig. 37.

affect it. By this means the image stamps an impression of itself upon the substance, but in this
impression the bright parts of the image appear dark
and the dark parts bright, and it is therefore called a
negative. From this negative the ordinary pictures
or **positives** are afterwards taken.

72. **Magnifying glasses.**—A lens may be used
for magnifying anything very small, thus forming a
magnifying glass with which most of you are no doubt
familiar. In this case you must place the glass very
near the thing that you wish to magnify. For instance,
you could not by means of a magnifying glass of this
kind magnify a distant object such as a planet or the

moon, but you can only magnify something close to you. If you wish to magnify a planet or the moon, you must use two glasses ; one a large glass by means of which you get an image of the planet or of the moon—just as by means of a burning glass you get an image of the sun—and the other a magnifying glass, by means of which you examine and enlarge this image, which the other glass has given you.

Thus, if you wish to magnify a near object you use a magnifier, but if you wish to magnify a distant object you must first of all, by means of a lens, obtain near at hand an image of the distant object, and then treating this image just as you would the object itself, you may scrutinise and magnify it by means of a magnifying glass. This combination of two glasses, one giving you an image of a distant object, and the other magnifying this image, is called a **telescope** ; in practice the glasses are shut up in tubes so as to keep out stray light.

73. **Different kinds of Light are differently bent.**—I have shown you how a ray of light is bent in passing through a prism. I have now to tell you that this bending is not the same for every kind of light. In fig. 38 we see how a ray of red light is bent after passing through a prism. If the ray had been orange instead of red, it would have been somewhat more bent out of its original course ; if yellow, still more ; if green still more than the yellow ; if blue still more than the green ; if indigo, more than the blue ; and if violet, still more than the indigo. Now if the

ray were a compound ray containing mixed together
all these seven colours (red, orange, yellow, green, blue,
indigo, and violet), each of these as it came out of the
prism would be bent differently from its neighbours,
and would therefore be separated from them, and
the eye would therefore see all these colours separate,
although they were mixed together when they
entered the prism,

A prism thus breaks up a compound ray of light into
its elements, separating the various colours from one
another.

Now you will perhaps be surprised when I tell you
that white light, such as that of the sun, is in reality
composed of a mixture of all the various colours which
I have given you above—red, orange, yellow, and so
on ; a little reflection will, however, convince you that
such is really the case.

We are all of us familiar with the magnificent dis-
play of colours seen in drops of dew, in crystals and
in gems, when rays of light are allowed to fall upon
them.

On such occasions they sparkle with all the colours
of the rainbow, and this very allusion bids us ask if
the hues of the rainbow be not due to the same cause
as the colours of gems. Does not its very name imply
the presence in the sky of a multitude of minute drops
of water such as would shine forth in the grass like in-
numerable diamonds ? Are not all these displays due
to the same cause ; and, if so, what is the cause ? The
discovery of it is due to Sir Isaac Newton, who was

the first to show that white light is in reality composed
of a great many differently coloured rays mixed together, and that these rays are in their passage through
certain substances separated from one another. The
prism, in fact, as we have already said, gives us the
means of separating the variously coloured elements of
a compound ray from one another.

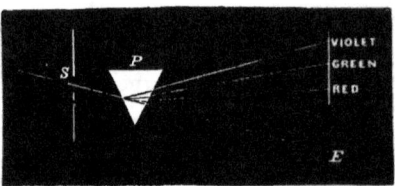

Fig. 38.

Suppose, for instance, that we have a narrow
vertical or up-and-down slit in the shutter of a darkened
room through which the full sunlight is allowed to
pass ;—in fig. 38 we have a plan of this arrangement
looking down upon it from above, or taking, as it were,
a bird's-eye view of it. Now if we have no prism to
commence with, and look from E towards the slit in
the shutter at s, we shall see a bright slit and nothing
more ; in fact, the slit will serve as an opening through
which we may see the bright sun beyond. Let us now
introduce the prism as in the figure ; when we have
done so, our eye at E will no longer see the slit. If,
however, we move our eye towards the thick part of

the prism, we shall at last catch hold of the light from
the slit, but it will be now very much changed in
appearance. It will not now reach our eye in the
shape of a bright thin slit, as formerly, but it will
appear as a broad band or ribbon of light of many
colours, beginning with red at the one end, and
passing gradually and in order through orange,
yellow, green, blue, and indigo, into violet at the
other extremity.

All this may be easily explained by what we have
already said, bearing in mind that white sunlight is in
reality composed of all the different colours mixed
together. Not only, therefore, are the rays bent in
their passage through the prism, but they are un-
equally bent. And we shall have for each variety of
light its own appropriate slit in its own appropriate
position. We shall therefore have a multitude of
little bright images of the slit lying side by side,
forming, in fact, a band or ribbon of light rather than
a slit ; the red being at one end, because the red
rays are least bent, and the violet at the other end,
because the violet rays are most bent. This variously
coloured ribbon of light is called a **spectrum** ; and if
it be the light of the sun which we employ to light up
our slit, then we get the **solar spectrum.**

74. **Recapitulation.**—I have now told you a good
deal about radiant light and heat. You have in the
first place learned that, as you begin to heat bodies,
they give out first of all dark rays, but that as you
raise their temperature, the rays become luminous and

capable of affecting the eye. You have then been
told something about the reflexion of these rays from
polished surfaces. You have also been told how their
direction is bent when they pass through water and
glass; and how a glass prism bends the rays towards
its thickest part. You have next been told that a
lens bends the rays all round towards its centre or
thickest part; and how, if you allow sun-light to fall
upon a lens, you get a small bright image of the
sun; which image will set fire to a sheet of paper or
burn the hand.

You have also learned that the moon or a planet
will give by means of a lens an image of the same
kind; and how, if you approach such an image with a
magnifying glass and look into it, you really see a very
large moon or a very large planet, and that this combination of two glasses is called a telescope. Finally,
you have been told that differently coloured rays of
light are bent by a prism into different places, so that
a prism separates all the elements of a compound ray
of light.

And now, before concluding, let us study a little
the nature of heat.

75. **Nature of Heat.**—We have already compared
heat to sound, and told you that a heated body is an
energetic body. Let us now take up this comparison
once more. In sound we have two things to study :
first, the body which vibrates; and secondly, the im-
pulses which this body sends through the air to our
ear, and which make us hear a sound.

Now you were told that a heated body is one in which the small particles are in very rapid vibration, and that just as a vibrating body gives out sound, which strikes the ear, so a heated body gives out light, which strikes the eye. But how is a body made to vibrate; a bell or a drum, for instance?—only by giving it a blow. You bring the heavy hammer or tongue quickly against the side of the bell, and the bell begins to vibrate; now this hammer or tongue before it strikes the bell is a body in rapid motion, and therefore possesses energy, or can do work. Well, what becomes of its energy after it strikes the bell? It has, in truth, given up its own energy to the bell, for the bell is now vibrating, and you have already been told that a vibrating body is one with energy in it. Thus the energy of the blow given to the bell has not been lost, but has only been transferred from the hammer to the bell. Now let us suppose a blacksmith places a piece of lead upon his anvil and brings down his hammer upon it with a heavy blow. You hear a dull thud, but there is no vibration like that of the bell. What becomes therefore of the energy of the blow? It is not transformed into vibrations like those of the bell, which can strike the ear—into what therefore is it changed? or is it changed into anything? We reply that it is changed into heat. The blow has heated the lead and set all its particles vibrating, although not in the same way as those of the bell; and if the blacksmith strikes the piece of lead long enough, I dare say he will even melt the lead.

No doubt some of you have spent much energy in rubbing a bright button on a piece of wood. Now what has become of all the energy you have spent upon the button ? We reply, it has been transformed into heat, as you will easily find out by putting the button quickly on the back of your own hand or on the back of your neighbour's.

EXPERIMENT 52.—To show you how the energy of a blow is changed into that other kind of energy which we call heat, let us take one of those wax matches tipped with phosphorus called vestas, and, placing it upon the hearthstone, strike it a blow with a hammer or stone ; you will now find that the heat developed has been sufficient to set the phosphorus on fire.

You thus see that friction produces heat, and you may have noticed that on a dark night sparks fly out from the break-wheel, which stops the motion of a railway train. In all such cases, actual visible energy is being changed into that form of energy which we call heat, the difference being that in visible energy the body moves as a whole, and all its particles move in the same direction at the same moment, while in heat the various particles move backwards and forwards rapidly, while the body, as a whole, is at rest. You thus see that visible energy can be changed into heat, and I have further to tell you that heat can to some extent be transformed back into visible energy. In the case of a steam-engine what is it that does all the work ? Is it not the fire that heats the water of the boiler? and in this case part of the heat-energy of the burning coals actually and truly changes itself into the visible energy with which the piston moves up and down, and the fly-wheel moves round and round.

In fact, all the work done by steam-engines is work got out of heat. Thus you see we can not only change actual energy into heat, but, in the steam-engine, we can change heat back again into actual energy.

ELECTRIFIED BODIES.

76. Conductors and Non-conductors.—It was known more than two thousand years ago that when a piece of amber is rubbed with silk, it attracts light bodies; and Dr. Gilbert, about three hundred years ago, showed that many other things, such as sulphur, sealing-wax and glass have the same property as amber.

Here you see the faint and small beginning of our knowledge of **electricity,** a knowledge which has of late years grown so wonderfully as to enable us to send messages between Europe and America in less than one second of time.

* EXPERIMENT 53.—Let us take a metal rod, having a glass stem, and rub the glass with a piece of silk, both silk and glass being well heated and quite dry. The glass will now have the power of

attracting little bits of paper or elder pith, but only at that place where it has been rubbed. The glass has, in fact, by rubbing, acquired a new property, but this property cannot spread itself over its surface. So much for glass. Suppose now that we take the metal rod and touch with it the prime conductor of an electric machine in action, we shall find that the metal rod has acquired the same properties as the glass ; that is to say, it will attract light bodies like paper or elder pith, but all parts of the rod of metal will have the same property, and not merely that part which touched the electric machine. In fact, the electric influence can spread itself over a surface of metal, though it cannot over one of glass. Glass, therefore, is said to be a **non-conductor** of electricity, while metal is called a **conductor**. In fact, neither heat nor electricity can easily spread itself over glass, but both can easily spread themselves over metal ; charcoal, acids, soluble salts, water and the bodies of animals are likewise good conductors of electricity, although not so good as the metals ; while, on the other hand, india-rubber, dry air, silk, glass, wax, sulphur, amber, shellac, are all very bad conductors.

If we wish to succeed in experiments with electricity, it is absolutely necessary to keep the electricity once we have got it ; we must, in fact, surround it on all sides by non-conducting bodies. It is, therefore, of great importance to make our experiments in dry air, and to make the body which has the electricity stand upon a glass support.

77. **Two kinds of Electricity.** EXPERIMENT 54.—I have now to convince you that there are two opposite kinds of electricity. To prove this let us make use of the apparatus you see in fig. 39, consisting of a small pith ball suspended by means of a silk thread to a glass support. First of all let us rub a glass rod with silk, and with the rod so rubbed touch the pith ball. The glass rod will communicate electricity to the pith ball, and it will not be able to get away, because the silk thread, the glass support, and the air (if dry) around the pith ball are all non-conductors. Now, if you notice, you will see that after the glass rod has been made to touch the pith ball, this ball will no longer be attracted to the glass rod, but will, on the other hand, be repelled by it. Let us next rub a stick of sealing-wax with a piece of warm, dry flannel, and bring the stick so rubbed near to the pith ball. It will now be found that the pith ball, which was repelled by the excited glass, will be attracted to the excited sealing-wax.

It thus appears that a pith ball, first touched with excited glass will be afterwards repelled by excited glass, but will be attracted by excited sealing-wax.

Now if we had reversed our plan of operations, and had first of all touched the pith ball with excited sealing-wax, instead of excited glass, it would then have been repelled by excited sealing-wax, but attracted by excited glass.

We learn from this that there are two kinds of electricity ; namely, that which we get from excited glass, and that which we get from excited sealing-wax.

Now when we touched the pith ball with excited glass, we communicated to it part of the electricity of the glass; and as it was afterwards repelled by excited glass, we conclude that **bodies charged with the same kind of electricity repel one**

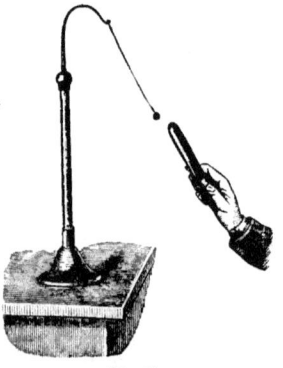

Fig. 39.

another. On the other hand, the pith ball, if charged with excited glass, will be attracted to excited sealing wax; or if charged with excited sealing-wax, it will be attracted to excited glass, and hence we conclude that **bodies charged with opposite kinds of electricity attract one another.**

78. **They exist combined in unexcited bodies.**—We may suppose that every substance has in it a quantity of these two kinds of electricity mixed together, and that what we do in rubbing is merely to separate the two electricities from one another. Accordingly, when we rub a piece of sealing-wax with a piece of flannel, all that we do is to separate the two kinds of electricity—the one kind keeping to the sealing-wax, while the other remains behind upon the flannel. In like manner all that we do when we excite glass with silk is to separate the two electricities, one remaining on the glass while the other adheres to the silk. The same thing holds in all cases where electricity is developed by friction, and it is impossible to produce the one electricity without, at the same time, producing just as much of the other also. In fine, we do not create electricity; but, according to this view of it, we merely separate the two opposite kinds from one another.

The electricity which appears in a stick of glass when it has been rubbed with silk is called **positive,** and that which appears in a stick of sealing-wax, when it has been rubbed with flannel, is called **negative.** These are merely terms used in order to distinguish between the two kinds of electricity.

79. **Action of excited on unexcited bodies.** —We have seen that electricities of the same kind repel, while electricities of opposite kinds attract each other, but we have still to learn what will happen in the following case. Let A (fig. 40), be a large ball of

E

hollow brass, and let the tube to the left hand of it be also of brass; also let these stand upon a glass support, so that any electricity which A has may not be able to get away.

Now let B and C be two vessels having their upper parts made of brass, only capable of being separated from one another at the middle part, where you see the line in the figure; and let both B

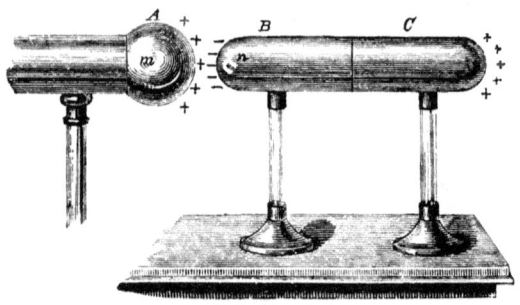

Fig. 40.

and C stand upon glass supports, so that any electricity which either of them has, may not be able to get away.

Let us begin by supposing that A has received a charge of positive electricity, and that in the meantime B and C are unelectrified. Now push B and C up towards A. Since B and C are not electrified, their two electricities are not separated from each other, but mixed together; however, when you push them up to A, the positive electricity of A attracts the negative electricity of B to its side, and repels the positive away to the extreme right of C, as you see in the figure.

If we now pull C away from B, and finally pull B away from A, we shall thus have got a quantity of negative electricity in B, and a quantity of positive in C, both separate from each other, while the electricity in A will be the same as before.

We have, in fact, made use of the electricity in A to separate part of the two electricities of B and C from each other, and A is still as ready as ever to help us again. Now this distant action or help, rendered by the electricity of A in separating that of B and C, is called **electric induction**.

80. **The Electric Spark.**—We may, however, perform our experiment in a somewhat different manner. Let us now bring B and C slowly towards A, and continue to do so. When A and B are very near together, we shall have the positive electricity of A and the negative electricity, which has been induced to appear on B, separated from each other by only a small thickness of air until at last they will be so strong and the film of air so thin, that they will rush together and unite in the form of a spark. The consequence will be that A will have lost a portion of its positive electricity, and B will have lost its negative. If we now pull B and C away there will still be the positive charge at C, which has not gone away; in fact, while A has lost part of its positive electricity, C will have gained just as much, so that

the result is virtually the same as if part of the electricity of A had gone over to C.

81. Sundry experiments.—What we have now said about electric induction may be easily illustrated

by a few simple and striking experiments; but it must be remembered that in all these experiments the glass of the apparatus must be quite dry and warm.

EXPERIMENT 55.—Here you see in the figure an instrument by which we can detect electricity, called the **gold leaf electroscope**. In order to show you its action, let me first of all communicate to the

Fig. 41.

knob at the top (see Appendix) a slight charge of positive electricity. Now this charge runs to the gold leaves which are electrically connected with the knob, and then these leaves, being both charged with the same kind of electricity, begin to repel each other as you see in the figure. The electroscope is now in action.

EXPERIMENT 56.—Having thus charged the electroscope with positive electricity, let us bring near its knob an excited glass rod, when the gold leaves will diverge still more. The reason of this is that the positive electricity of the excited glass decomposes the neutral electricity of the knob attracting the negative to itself, and repelling the positive to the gold leaves. If, therefore, the leaves had been previously charged with positive electricity, they will now diverge more widely.

EXPERIMENT 57.—If we now bring near the knob of the electroscope, charged as before with positive electricity, a stick of excited sealing-wax, we shall first find that the gold leaves will collapse instead of diverging. The reason is that the negative electricity of the excited sealing-wax decomposes the neutral electricity of the knob attracting the positive to itself, and driving the negative to the gold leaves. But since the gold leaves were previously charged with positive electricity, part of this charge will be cancelled by the negative electricity driven towards them, and they will consequently collapse.

EXPERIMENT 58.—Here we have a hollow brass ball or conductor, supported on an insulating glass stand. Let us now bring this insulated conductor near the electric machine when in action, and we shall get a spark, but it will be very feeble. Let us now touch with our finger that part of the hollow ball which is farthest from the machine, and the spark given to the ball will now be much more intense.

This illustrates what we said in Art. 80 about the cause of the spark. In fact, the positive electricity of the machine pulls towards itself the negative electricity of the hollow ball, and drives away the positive as far as possible. If, however, this ball is insulated, the positive cannot be driven away sufficiently far, nor the two electricities be separated sufficiently well, and the consequence is you have but a feeble spark. But when you touch the hollow brass ball, the positive electricity of the ball is driven through your body to the earth, the electricities are thus well separated, and there is a good spark.

82. Action of Points.—In the last experiment, if you continue to touch the brass ball, and the electric machine is worked at the same time, a succession of sparks will pass through your body to the earth, and these will cause you to feel a somewhat unpleasant sensation. The spark from the electric machine may in truth be compared to a flash of lightning—a flash of lightning being, in fact, a very long spark. Now, just as when a man is struck by lightning the electricity passes through his body to the earth, so when we grasp or touch the ball of the last experiment, the electricity goes through our body to the earth.

EXPERIMENT 59.—Now let us attach a point to the hollow ball, and place this point next the conductor of the machine, touching the ball as before with our finger. It will now be impossible to get a spark from the machine, but there will be instead a continuous rush of electricity. In fact, anything pointed carries off the electricity just as rapidly as it is produced, and does not give it time to gather so as to form a spark. We now see the use of the pointed metallic conductors that are placed above lofty buildings, to protect them from lightning strokes. These pointed metallic conductors, running down into the earth, carry off the electricity in a silent manner, just as the point did in Experiment 58 ; and just as the point protected my finger from a spark in the one case, so does the lightning conductor protect the building from a flash or stroke of lightning in the other.

Franklin, an American philosopher, was the first to find out that lightning and electricity are the same thing—the only difference being that a flash of light- ning is often several miles in length, whereas an electric spark is only a few inches.

83. **Electrical Machine.**—You are now in a position to understand the construction of an electric machine. Such a machine is composed of two parts ; we have first of all an arrangement for producing electricity, and we have next an arrangement for collecting it.

One of the best known machines is that in which the electricity is produced by a large plate of glass revolving, as in fig. 42. As the plate of glass revolves, it is rubbed against by two sets of rubbers, one above and the other below. These rubbers are usually made of leather stuffed with horse-hair, so as to press rather tightly against the glass. They are coated with a soft metal, which is spread over the leather, and this metal is generally made of one part of zinc, one of tin, and two of mercury, melted together. There is a metallic chain which connects these rubbers with one another, and with the earth. Now when the glass plate is turned round, positive

electricity is produced in the glass, while negative
electricity is produced in the rubbers. The negative
electricity of the rubbers then passes along the
metallic chain which is connected with the rubbers,
and is conducted by means of this chain to the earth,
through which it spreads until it is scattered and
diffused—in fact completely lost. We have thus got

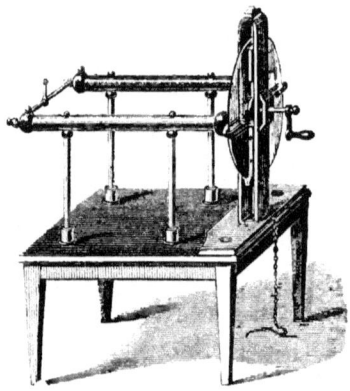

Fig. 42.

rid of the negative electricity, and there is now left
the positive electricity on the glass. Now surrounding the glass we have two brass rods, which are united
to a large metallic surface called the **conductor**,
which you see in the figure. This conductor stands
upon glass supports, so that it is able to keep what
electricity it gets. The two large rods near the glass
plate are moreover armed with metallic points. Now

you have already been told that points have a
great tendency to draw off electricity. The consequence is that these points draw off, or collect,
the positive electricity of the glass and carry it
to the conductor, where it remains, since the conductor stands upon glass supports. By turning the
glass plate sufficiently long, we may thus accumulate a large amount of positive electricity in this
conductor.

EXPERIMENT 60.—If, when the conductor of the
electric machine is charged with electricity, I place
my finger near it, a spark passes between the conductor and my finger. The reason is that the positive
electricity of the conductor separates the two electricities which are together in my finger, driving
away the positive, which is of the same kind as itself,
to the earth through my feet, but, on the other hand,
attracting the negative to itself.

The two electricities—namely the positive in the
conductor, and the negative in my finger—then rush
together through the air and unite with each other,
and in so doing they form a spark.

84. Leyden Jar. EXPERIMENT 61.—When you
thus approach your finger or your knuckle to an
electric machine, you feel a pricking sensation when
the spark passes, but that is all; you do not get a
severe shock. In order to get a shock you must
use a Leyden jar, such as you see in fig. 43. This
is a glass jar, the inside of which is coated with
tinfoil, as well as the outside up to the neck. A

brass rod with a knob at the end is connected with
the inside coating, and is kept tight by being passed
through a cork which covers the mouth of the jar.
Thus the jar has two coatings, an inside and an
outside one, and these are quite separated from each
other, as far as electricity is concerned, inasmuch as
glass does not conduct electricity.   Now suppose I
take the jar by its outside coating in my hand, and

hold the inside knob to the
conductor of an electrical
machine at work.   The
positive electricity from the
conductor will then get into
the inside coating of the
jar.   It will then decompose the two electricities of
the outside coating, driving
away the positive through
my hand and body generally

Fig. 43.

to the earth, and attracting the negative.   In fact
there will be a battalion of positive electricity in the
inside coating facing an opposite battalion of negative
electricity in the outside coating, the two longing very
much to meet, but unable to do so for the glass.   So
intent are these two electricities on watching each
other that they will remain close at their post while I
put some more positive electricity into the interior.
This second charge will then act precisely like the
first ; it will decompose anew the two electricities of
the outside coating, driving positive electricity from

the outside through my hand to the earth, while negative electricity will remain in the outside coating,
facing the new battalion of positive electricity which
has been introduced inside.   We have now two inside
and two outside battalions watching one another, and
by continuing this process we accumulate a large
quantity of opposite electricities in the two coatings
of such a jar.

If we wish to discharge the jar, we make use of a
discharging rod, such as you see in the figure.   It
should be held by its glass handles, and one of the
knobs should be made to touch the outer coating of
the jar while the other is gradually brought near the
knob connected with the interior of the jar; when
the two knobs are near together a bright spark is
seen, accompanied with a
report, and the jar is discharged.   If we wish to
feel the shock ourselves, let
us grasp the outside coating
by one of our hands, and
approach the other towards
the knob connected with
the inside coating, the dis-

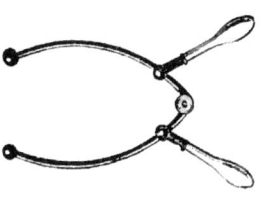

Fig. 44.

charge will then take place through our body.   Or if
several wish to feel the shock, let them all join
hands, and let the one at the one end grasp the outside coating, while the one at the other end touches
the inside knob, and the shock will then pass through
the bodies of all.

85. **Energetic nature of electrified bodies.**— From what has been said you must now be convinced that electricity is something which has energy in it. You see that the two opposite electricities of the jar rush together and unite, and that the union is accompanied by a flash and a report. This flash is very bright while it lasts; and although it does not last longer than the twenty-four thousandth part of a second, it nevertheless implies considerable heat. Now heat means energy, and we thus see that when a jar is discharged, that kind of energy which we call electricity is changed into that other form of energy which we call heat and light.

Again, since electricity is an energetic thing, it requires labour or work to produce it; you do so by turning the electric machine, but such a machine is particularly hard to turn on account of the electricity. You thus see that there is nothing for nothing; if you wish to obtain an energetic agent, you must spend work in doing so. On the other hand, there is no disappearance of energy when the two electricities combine, but only a change from the form of electricity into that of heat.

86. **Electric Currents.**—You have seen that when you hold a pointed conductor near an electric machine at work (Art. 82) there is a continuous stream or current of electricity, which passes through the point and through your hand to the ground.

We have, however, a much better means than the electric machine gives us of obtaining powerful electric currents. We shall now briefly describe this method, which was first discovered by an Italian called Volta, and which has therefore been named the **Voltaic battery.** This arrangement is shown in the figure below. Here you see to the extreme left a plate marked c, which means a plate of copper. Next you see a plate of zinc marked z, which is soldered to a wire connecting it with the plate of copper in the

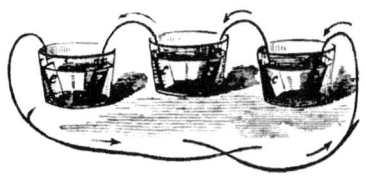

Fig. 45.

second vessel. In the second vessel you have another plate of zinc, which is similarly connected with the copper in the third vessel. Finally, to the extreme right you have a single plate of zinc. Suppose now that the vessels are filled with a mixture of sulphuric acid and water, and that we attach wires to the copper at the left-hand end, and also to the zinc at the right-hand end, and that we bring these wires together. (These wires are called the pole-wires of the battery.) It will now be found that there is a current of positive electricity passing round and round through the circuit in the direction of the arrow-heads. Let us trace how it goes. In the first place, it comes from the wire attached to the extreme left-hand copper

plate, and goes, as in the figure, through the long
wires until it enters the extreme right-hand plate of
zinc; it then passes through the liquid till it reaches
the copper plate, from which it passes along the wire
to the next zinc plate; it then passes through the
liquid of the middle vessel to the copper plate, and
from that by the wire to the zinc plate of the left-
hand vessel; and finally from the zinc plate of this
vessel through the liquid to the same plate from which
it started originally.

87. **Grove's Battery.**—The arrangement now
described was that used by Volta, but since his time
many improvements have been made in the method
of obtaining a current of electricity.

It was found that with Volta's arrangement the
current, though strong at first, very soon became
weak; but a method has been devised by which the
electric current can always be kept at the same
strength. A battery by which this is done is called
a constant battery, and one of the best is that
invented by Grove (see fig. 48). In this battery,
instead of a single vessel we use a double one, the
outer vessel being made of glass, while the inner is
made of porous earthenware. The outer glass or
stoneware vessel is partly filled with diluted sulphuric
acid. Within it we have a plate of zinc (amalgamated
on the outside), as you see in the figure, while within
the glass vessel we have a porous vessel, made of
unglazed porcelain. Into this porous vessel is poured
strong nitric acid, and into this nitric acid is put a

thin plate of platinum, which takes the place of the copper in Volta's arrangement.

Now when this battery is in action, the zinc dissolves in the dilute sulphuric acid, and during this process hydrogen gas is given off. But this hydrogen does not rise up in the shape of bubbles, but finds its way into the porous vessel which contains the strong nitric acid. It there decomposes the nitric acid, taking some oxygen to itself, so as to become water (hydrogen and oxygen forming water), and thereby turning the nitric into nitrous acid, which shows its presence by strong orange-coloured fumes. Thus the hydrogen does not reach the platinum plate ; indeed it is to prevent its doing so that this arrangement was made, for it was found that in Volta's original battery the hydrogen given out as the zinc dissolved adhered to the copper plate, in consequence of which the force of the battery became much weakened.

What we have now described is only a single vessel, or cell, as it is called, of Grove's battery. In a large battery of this kind there may be 50 or 100 cells— the wire that is attached to the platinum of one cell being connected with the zinc of another, in a manner precisely similar to that of fig. 45, the only difference being that instead of copper we have platinum, and instead of a single vessel a double one of the nature now described. Also, the positive current passes through the liquid from the zinc to the platinum plate, just as it passed through the liquid from the zinc to the copper plate in Volta's arrangement.

88. Properties of the Current.—Let us now see what an electric current can do; that is to say, let us perform a few simple experiments.

EXPERIMENT 62.—Make a Grove's battery ready for action, and introduce a bit of very fine platinum wire between the two pole wires of the battery; when the connexion is made, and the current passes, it will be found that the fine platinum wire will become red-hot.

EXPERIMENT 63.—Make a Grove's battery ready for action, and insert its two pole wires into two inverted vessels containing water, as in fig. 46. It

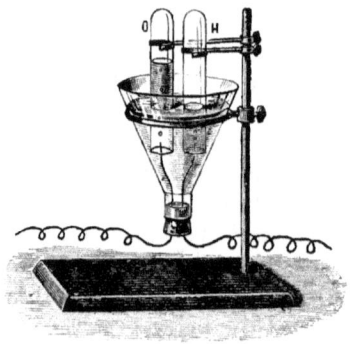

Fig. 46.

will be found that the current decomposes the water and that oxygen gas will appear in the one vessel and hydrogen gas in the other. The oxygen gas will appear at the pole which is connected with the platinum plate, while the hydrogen will appear at that which is connected with the zinc plate. Thus you see that a voltaic battery has the power of decomposing water. It has also the power of decomposing very many compound liquids.

EXPERIMENT 64.—Here we have some copper wire covered with thread so as to insulate it, and this copper wire is wound round a thick piece of iron shaped like a horse-shoe ; now let us connect the two poles of our battery with the two extremities of the copper wire which goes round the iron. If the battery be now in action, it will be found that the iron has acquired the power of attracting other iron towards it, so that a plate of iron will be held up, as in the figure, with a heavy weight

Fig. 47.

attached to it. As soon, however, as the connexion between the horse-shoe and the battery is broken this power is lost, and the weight which the iron has been supporting will drop down at once.

EXPERIMENT 65.—Take a bit of hard steel, such as a knitting needle, and attach it to the iron of the horse-shoe in the last experiment while the current is passing. This needle will have gained certain properties which (unlike the soft iron) it will not lose when the current is broken, but will retain ever afterwards. For instance, if we suspend the needle round the middle by means of a very fine silk thread, and let it swing horizontally, it will always point in one direction, and this direction will be nearly north and south. The needle will, in fact, have become a compass needle, always pointing in one direction, and thus enabling the mariner when out at sea to steer his vessel in the proper course. A piece of hard steel possessing these properties is called a magnet.

EXPERIMENT 66.—Let us now suspend a magnetic needle horizontally upon a pivot. It will point nearly north and south. But let us bring near it a wire through which a current is passing, and it will be found that the needle will no longer point north and south, but it will place itself so as to lie across or at right angles to the wire conveying the current.

If we now break the current, the needle will resume its usual direction.

EXPERIMENT 67.—We may render the last experiment more marked by means of an arrangement such

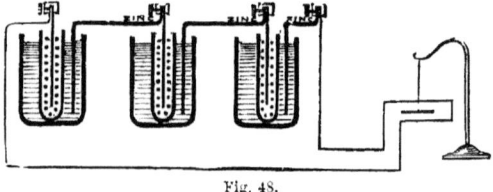

Fig. 48.

as is sketched in the above figure. Let us suppose that we have our battery at one end of the room, while two wires covered with thread are carried from

the two poles of the battery quite to the other end of
the room, and are there joined together so that
the battery is now in action. Furthermore you see
at the end most remote from the battery a suspended
magnetic needle, which is placed near the wire, and
which will be violently deflected when the current
passes. Now if any one at the very opposite corner of
the room should disconnect the wire from one of the
poles of the battery, that very moment the current
will cease to flow, and the magnetic needle will
resume its ordinary position.

89. **Electric Telegraph.**—It thus appears that
by disconnecting the wire from the battery at one end
of the room the needle is made to move at the other
at the very same moment. This action would take
place even if the wires connected with the poles were
carried 100 or even 1,000 miles away before they
were joined together. If a magnetic needle were
placed beside the wire conveying the current, even
though the wire should be 1,000 miles from the
battery, it would be deflected, but as soon as the other
extremity of this wire 1,000 miles away was disjoined
from the pole of the battery, the current would cease
to pass, and the magnetic needle would return to its
usual position. **You thus see how it is possible,
by making and breaking contact of a wire with
the pole of a battery, to move a magnetic
needle 1,000 miles away.**

In fact we have here the principle of the electric
telegraph, which performs such wonders in the way

of information, telling us what takes place in America a few seconds after it happens. I cannot, however, enter more fully into the subject, but at least you see that it is possible to agitate a magnetic needle 1,000 miles away, and, just as in the alphabet for the deaf and dumb, these signals may be made the means of conveying information.

90. **Conclusion.**—You have now learned what the electric current can do. How, in the first place, it can heat a fine wire through which it passes; how, secondly, it can decompose water and other compounds; how, thirdly, it can make a piece of soft iron into a powerful though temporary magnet; how, fourthly, it can make a piece of hard steel into a permanent magnet; and fifthly and lastly, how it can deflect the compass needle, rendering it thereby possible to telegraph to great distances.

We cannot enter more fully into this very interesting subject, but in conclusion let me remind you that you have now learned something about the active moods of matter. We spoke first of all about moving bodies, then about vibrating bodies, then about heated bodies, and lastly about electrified bodies; and we have tried throughout to show you that the activity which a body may possess is never really lost. It may, no doubt, pass to some other body, or it may change its form, going from visible energy into sound, or into heat, or into electricity, or changing about in many different ways, but it is really lost no more than a particle of matter is lost.

Indeed just as the Science of Chemistry is built upon the principle that matter only changes form, going from one combination to another, and does not absolutely disappear, so the science of Physics is founded upon the principle that activity or energy only changes form and never absolutely disappears. This, however, is a principle the full development of which must be reserved for a future stage.

THINGS TO BE REMEMBERED.

A POUND avoirdupois is equal to 7,000 grains.

If a stone be dropped from the hand, it will fall through 16 feet during the first second of time.

Steel is the strongest metal, but gold is the most malleable; for a cubic inch of gold can be beaten out so as to cover the floor of a room 50 feet long and 40 feet wide.

The diamond is the hardest solid; that is to say, it can scratch everything else, but nothing else can scratch it.

A cubic inch of water weighs nearly 252 grains; and, therefore, four cubic inches weigh nearly 1,000 grains.

100 cubic inches of air weigh 31 grains.

100 cubic inches of carbonic acid weigh 47 grains.

100 cubic inches of hydrogen only weigh 2 grains.

The pressure of the atmosphere will support a

column of mercury 30 inches high, and a column of water more than 30 feet high.

Sound travels through air at a velocity of about 1,100 feet in one second of time.

If a musical string vibrates 50 times in one second, it emits a deep, low note ; if it vibrates 10,000 times in one second, it emits a shrill, high note.

The heat required to melt a pound of ice would heat 79 pounds of water one degree. The heat required to boil away a pound of boiling water would heat 537 pounds of water one degree.

Light travels through space nearly at the rate of 190,000 miles in one second of time.

The spark from a Leyden jar lasts only the twenty-four thousandth part of one second.

INSTRUCTIONS REGARDING APPARATUS.

The apparatus to be used should be set upon the table before the lesson, and the teacher should make sure that he can perform without difficulty the various experiments. After the lesson the apparatus ought to be put away carefully into its appropriate place.

Care must be taken that the piston of the air-pump is rendered tight in its cylinder by means of lard. Care must also be taken that the receiver fits well upon its bed-plate, and for this purpose it must be well greased with lard. When this is done, the receiver ought to move smoothly and without noise on

its bed-plate ; but if there is a grinding noise it shows
that some hard substance is present, and the bottom
of the receiver must then be carefully cleaned and
greased anew. This remark applies to the hemispheres
(fig. 15), as well as to the glass receivers.

In order to fill the box of Experiment 28 with carbonic
acid gas, the tube conveying the gas should *descend*
very nearly, but not quite, to the bottom of the box.

To fill the same box (Experiment 29) with hydrogen,
the tube conveying the gas should *ascend* very nearly
to the bottom of the box, which is now uppermost.

The whole apparatus for Experiment 45 must be
placed in a cool room some hours before the experiment is made.

Great care ought to be taken in handling phosphorus, which very easily takes fire. The stock of
phosphorus should be kept under water, and the little
bits cut off should be well dried in blotting-paper
before being used.

When the mercury is tarnished, take a piece of
paper and make it into a funnel, having a pin-hole at
its bottom. Pour the mercury gently into this funnel,
and let it run through the pin-hole into a vessel pre-
pared for it. It will now be quite bright.

Care must be taken that the mercury is not con-
taminated with other metals. A small portion should
be kept separate for amalgamation in the battery.

Before the electrical machine is used the glass plate
ought to be well warmed. For this purpose it ought
to be placed endwise towards the fire, and the handle

ought to be turned round occasionally, so as to expose
to the fire the various parts of the plate. If these
instructions be not attended to, the glass may probably
crack.

The electroscope ought not to be charged highly,
otherwise the gold leaves will be driven to the sides
of the jar and be torn. To charge the electroscope
give the Leyden jar a single small spark from the
machine—then touch the electroscope with its knob.

The insulating glass supports of the conductor
ought also to be warm and dry.

Finally, the Leyden jar and everything made of
glass with which any electrical experiment is to be
made, ought to be warm and dry.

In the Grove's battery the zinc ought to be well
amalgamated (see Chemical Primer), and the various
metals ought to be quite bright at the points where
they are connected with the battery.

The outer cells ought to be charged with one part
by measure of strong sulphuric acid and eight parts
of water.

The porous vessels of the Grove's battery ought to
be well steeped in water after the battery has been in
use ; and the zinc and platinum plates ought likewise
to be well cleaned.

In Experiment No. 66, it is necessary to fill with
mercury the two little brass cups into which the ends
of the battery wires are plunged.

QUESTIONS.

INTRODUCTION.

I. Definition of Physics.—1. Give an instance of two different *kinds* of things.

2. Give an instance of two different *affections* or *moods* of the same thing.

II. Definition of Motion.—1. Before you understand a motion you must know two things about it. What are these?

2. One man going always at the same pace walks eight miles in two and a quarter hours while another walks four miles in one hour, which walks fastest?

3. A man walks ten miles in two and a half hours. What is his rate of motion? A cannon ball moves over 6,600 feet in 5½ seconds. What is its rate of motion?

III. Definition of Force.—1. What do we mean by force?

2. Give an experimental instance of a force which produces motion in a body previously at rest.

3. Give an experimental instance of a force which stops a moving body.

4. Give an experimental instance of a force which is prevented from acting by another force.

THE CHIEF FORCES OF NATURE, p. 16.

I. Definition of Gravity.—1. What is the cause of the weight which things have?

2. Suppose you could annihilate the interior of the Earth (preserving its crust to stand on), would there be any alteration in the weight of a lump of lead?

3. Suppose you could hold a pound of lead in your hand in the middle of empty space without the Earth being under you, would the lead have weight?

II. Definition of Cohesion.—1. Give an instance of cohesion.

2. What is the most characteristic difference between gravity and cohesion? Illustrate your reply by an example.

III. Definition of Chemical Attraction.—1. Give an instance of chemical attraction.

2. What is the peculiar characteristic of this force?

PHYSICS.]

IV. Use of these Forces.—1. What would happen if there were no gravity?

2. What would happen if there were no cohesion?

3. What would happen if there were no chemical attraction?

HOW GRAVITY ACTS, p. 21.

I. Centre of Gravity.—1. What do we mean by the centre of gravity of a body?

2. Has every substance a centre of gravity?

3. If a body be free to move, how will it place its centre of gravity?

4. Describe a practical method of finding the centre of gravity of an irregular plane sheet of heavy matter.

5. Could this method be practically followed if the sheet were not all in one plane? Give a reason for your reply.

II. The Balance.—1. Sketch the common balance.

2. Why could not the balance have its centre of gravity above the point of suspension upon which the balance is swung?

3. What makes the beam of a balance come back to a definite position when tilted aside?

THE THREE STATES OF MATTER, p. 24.

1. Name the three states of matter.

2. In which of these states does matter possess most cohesion? In which state has it no cohesion?

3. Describe an experiment showing that mercury has some cohesion.

4. Describe an experiment showing that water has some cohesion.

5. Define a solid.

6. Define a liquid.

7. Define a gas.

PROPERTIES OF SOLIDS, p. 27.

1. Is it *absolutely* impossible to alter the shape or size of a solid?

2. Enumerate the various ways in which you might try to break up or alter the form of a bar of iron.

3. Describe an experiment showing that the amount of bending of a beam is nearly proportional to the weight applied.

4. A weight of ten pounds applied as in Experiment 9 lowers the centre of a beam one eleventh of an inch. How much would the centre be lowered by a weight of twenty-eight pounds similarly applied?

PHYSICS.]

5. Describe an experiment showing that when a beam is so placed as to give it depth rather than surface, it is least bent by the application of a weight.

6. What is meant by the *limits of perfect recovery* of a solid structure?

7. What are the two essential points that an architect or constructor ought to attend to?

8. Define *friction* by means of an experiment.

9. What would happen if there were no friction?

PROPERTIES OF LIQUIDS, p. 32.

I. Size and Shape.—1. Does a liquid exhibit a strong tendency to retain its present shape?

2. Does a liquid exhibit a strong tendency to retain its present volume? Illustrate your reply by an experiment.

II. Liquids communicate Pressure.—1. Describe an experiment showing that liquids communicate pressure.

2. Describe an experiment showing that liquids communicate pressure in all directions.

3. Who discovered this last-mentioned property of liquids?

4. Describe an experiment showing that the pressure of a liquid against a piston is proportional to the area or surface of the piston.

5. Water presses against the surface of a square piston the side of which is two inches with a pressure of ten pounds, what will be its pressure against the surface of a piston similarly placed of which the side is three inches?

III. Water Press.—1. Sketch and describe the water press.

2. The area of the large piston of a water press is eighty times as great as that of the small piston. A force of fifteen pounds is communicated to the smaller piston. With what force will the large piston rise?

3. Will the large piston of a water press rise as fast as the small piston falls?

IV. Liquids find their Level.—1. Describe an experiment showing that the direction of gravity is perpendicular to a free surface of mercury or any other liquid.

2. Sketch and describe the Water Level.

V. Pressure of Deep Water.—1. Sketch and describe an experiment showing that the pressure of a liquid is proportional to the depth, and is exerted upwards as well as downwards.

PHYSICS.]

2. If the pressure against a surface be six pounds, ten feet below the surface of a lake, what will be the pressure against the same surface twenty-five feet below the surface?

3. Will the pressure at a given depth be different according to the size of the lake?

4. How could you illustrate this pressure by sinking a bottle in deep water?

VI. Buoyancy of Water.—1. Define by aid of an experiment the buoyancy of water.

2. Make an experiment showing that, while a substance *apparently* becomes lighter when weighed in water, yet there is no *absolute* loss of weight.

3. Make an experiment showing that, when anything is weighed in water it will suffer a loss of weight exactly equal to the weight of its own bulk of water.

4. Why will a piece of iron sink in water?

5. Why will a cork float in water?

6. When will a substance neither sink nor swim in water but remain at rest in any part of the liquid?

VII. Comparative Density.—1. What do you mean by the comparative density or *specific gravity* of a body?

2. A piece of pure gold weighs in air fifty-seven grains and in water fifty-four grains, find its specific gravity.

3. On what occasion and by whom was the discovery made of the method of determining specific gravities?

4. A piece of gold said to be pure weighs seventy-six grains in air and seventy grains in water. Is this gold pure? Give a reason for your reply.

5. A piece of stone weighs 200 grains in air, and 150 grains in water. Another piece of the same stone weighs 560 grains in air, what will it weigh in water?

VIII. Buoyancy of other Liquids.—1. Which has most buoyancy, a heavy or a light liquid?

2. Name a liquid in which iron will float.

3. Can a man float most readily in fresh water or salt?

4. Name a sheet of water in which a man will not easily sink.

IX. Capillarity.—1. Mention a case in which water will rise above its level.

2. Show by an experiment that this rising depends on the attraction of the water for the substance used.

3. Name a substance that has a similar attraction for mercury.

PROPERTIES OF GASES, p. 46.

I. Pressure and Weight of Air.—1. What is the characteristic distinction between a gas and a liquid ?

2. Whether is air repelled or attracted by the Earth ? Illustrate your reply by an experiment.

3. Describe an experiment showing that some gases are heavier bulk for bulk than air.

4. Describe an experiment showing that some gases are lighter bulk for bulk than air.

5. Does the ocean of air above us press against the Earth just as the ocean of water presses against the sea-bottom ?

6. Why is not a piece of paper pressed hard against the table by the weight of air above it ? Illustrate your reply by an experiment.

7. Describe an experiment showing that air has buoyancy.

II. The Barometer and its Uses.—1. Describe the barometer.

2. Who invented it ?

3. What is the usual height of the barometric mercurial column ?

4. Would this column be lengthened or shortened by carrying the barometer to the top of a lofty mountain ?

5. What is meant by the *Torricellian vacuum ?*

6. In what way does the height of the mercurial column in general vary with the weather ?

III. Air-pump.—1. What is meant by the words *piston, cylinder, valve ?*

2. Sketch an air-pump and describe its action.

3. The bell-jar of an air-pump contains 90 cubic inches, while the cylinder contains 10 cubic inches : what proportion of the air will be taken out of the bell-jar after one complete stroke of the piston ?

IV. Water-pump. Syphon.—1. If water instead of mercury were used for a barometer, would the column be longer or shorter ?

2. Approximately speaking, what would be the length of the column of a water barometer ?

3. Sketch the common water-pump and describe its action.

4. Why will not this pump work if the distance between the surface of water in the reservoir and the lower valve be greater than 30 feet ?

5. Why must the distance of question 4 be altered if the pump is worked on the top of a lofty mounta'n ?

6. Sometimes before using a pump it is necessary to throw a little water upon the piston. What is the object of this ?

7. Sketch a syphon and show how to use it.

MOVING BODIES, p. 60.

I. Energy and Work.—1. Is *Energy* a substance, or a *mood or affection* of a substance ?

2. What do we mean when we say a substance is full of energy ?

3. Enumerate the most conspicuous cases in which a substance is full of energy.

4. How do we measure energy ?

5. What is our *unit of work* ?

6. How much work will be done in raising $5\frac{1}{2}$ lbs. $10\frac{1}{2}$ feet high against gravity ?

7. A cannon pointed vertically upwards fires a ball weighing 200 lbs., which rises 850 feet before it turns. What is the energy of the ball ?

II. Work done by a Moving Body.—1. A stone weighing one pound projected upwards with the initial velocity of 32 feet per second will rise 16 feet ; how much energy does it contain ?

2. If a stone weighing four pounds be projected upwards with the velocity of last question, how high will it rise and how much energy will it contain ?

3. If a stone weighing three pounds be projected upwards with the (double) velocity of 64 feet per second, how high will it rise and how much energy will it contain ?

4. A cannon-ball discharged with the velocity of 1,000 feet per second will pierce through six planks of oak ; through how many similar planks will a similar ball pierce when discharged with the (double) velocity of 2,000 feet per second ?

III. Energy in Repose.—1. Is a lion when asleep or at rest totally devoid of energy ? If not, what kind of energy has he got ?

2. Give an instance showing that a pile of stones may possess energy on account of their position.

3. When is a reservoir of water possessed of energy ?

4. What is the kind of energy that a wind-mill makes use of ?

5. Specify the advantage which energy of repose has over active energy.

VIBRATING BODIES, p. 65.

I. Vibration—Sound.—1. Give an experimental instance of a moving body that does not change its place as a whole.

2. What is the name given to this peculiar species of motion ?
PHYSICS.]

3. Does a vibrating body give a series of blows to the air around it?

4. When this blow reaches our ears what do we call the sensation produced?

II. Noise and Music.—1. Give an instance of a body which deals a single blow to the air.

2. Give an instance of a body which deals a series of blows to the air.

3. What do we call the sensation produced when a single blow strikes the ear?

4. What do we call the sensation produced when a series of blows strike the ear?

5. What is the physical distinction between a deep low note and a shrill high note?

6. Give an instance showing that sound is a species of energy, and is capable of doing work.

III. Motion of Sound through Air.—1. Describe an experiment proving that sound requires air to carry it to the ear.

2. When a cannon gives a blow to the air, are the individual particles of air so struck shot off till they reach the ear of a man at a distance who hears the report?

3. If this be not the case, how is the motion propagated to his ear? Illustrate your reply by an experiment.

4. Give an illustration of this derived from the game of croquet.

IV. Its Rate of Motion.—1. Give a proof that sound requires time to go from a cannon to the ear.

2. At what rate does the sound pass through the air?

3. At what rate will sound pass through water?

4. At what rate will it pass through wood?

5. A man at a distance hears the report of a cannon five-and-a-half seconds after seeing the flash, how far is he from the cannon?

V. Reflexion of Sound—Echoes.—1. Give a physical explanation of echoes.

2. Describe an experiment showing that sound like light can have a focus.

3. Illustrate the property of sound by reference to a peculiarity of St. Paul's Cathedral in London.

VI. How to find the Rate of Vibration corresponding to a given note.—1. Sketch and describe an instrument by which we can find the number of vibrations in one second corresponding to any note.

PHYSICS 1

HEATED BODIES, p. 76.

I. Nature of Heat (first notice).—1. Is a hot body heavier than a cold one ?

2. Is a hot body possessed of more energy than a cold one ?

3. If heat be a species of motion, why does not the eye see the particles of a hot body moving ?

4. In *vibrating bodies* there are two things to be studied, what are these ?

5. In *heated bodies* there are two things to be studied, what are these ?

II. Expansion of Bodies when heated.—1. Describe an experiment showing that a metallic rod becomes longer when heated.

2. What happens when a hollow glass bulb filled with water is heated ?

3. What happens when a bladder two-thirds filled with air is heated ?

III. Thermometers, and how to make them.—1. Describe generally the instrument called a *mercurial thermometer* and its mode of action.

2. Describe the method of filling and sealing a mercurial thermometer.

3. Describe the method of graduating a *centigrade* mercurial thermometer.

4. Why is this instrument called a centigrade thermometer ?

5. What is blood heat on a centigrade scale ?

IV. Expansion of Solids, Liquids, and Gases.—1. whether does glass or lead expand most ?

2. Whether does platinum or zinc expand most ?

3. Give a proof, by means of the thermometer, that liquids expand more than solids.

4. Do liquids expand more or less rapidly at a high than at a low temperature ?

5. Do gases expand more than liquids ?

6. Do gases expand from any other cause than heat ?

7. If a bladder not completely filled with air have a volume $=1,000$ cubic inches at the freezing-point, what will be its volume at the boiling point ?

8. Describe an experiment showing that liquids expand with enormous force.

9. Show how the force of contraction due to cooling is made use of in making carriage wheels.

V. Specific Heat.—1. What is meant by the *specific heat* of a substance ?

2. Name a substance having a very great specific heat.

3. Name a substance having a very small specific heat.

4. Illustrate your replies to 2 and 3 by an experiment.

VI. Change of State.—1. In what order does the heating of a substance change its state ?

2. One piece of iron is white-hot but solid, another is melted ; which is hottest ?

3. One piece of iron has been melted and another driven into vapour, which has been heated most ?

4. Name a liquid that has never been frozen.

5. Name a gas that has never been liquefied.

6. Can we trust to the sense of touch in estimating temperature ?

7. What is meant by a *refractory* substance ?   Name one.

8. What is the melting-point of ice on the centigrade thermometer, and what the boiling-point of water ?

VII. Latent Heat of Water and Steam.—1. Define the latent heat of water by an experiment.

2. If a pound of ice at 0° C. be mixed with a pound of boiling water at 100° C., will the mean temperature be greater or less than 50° C. ?

3. Define the latent heat of steam by an experiment.

4. If a pound of ice-cold water at 0° C. be mixed with a pound of steam at 100° C., will the mean temperature be greater or less than 50° C. ?

5. What do we mean by saying that the latent heat of water is 79 ?

6. What do we mean by saying that the latent heat of steam is 537 ?

7. What would happen in certain countries if the latent heat of water were very small ?

8. What would happen if the latent heat of steam were very small ?

9. Describe an experiment showing that true steam is invisible.

VIII. Ebullition and Evaporation.—1. State the difference between *ebullition* and *evaporation*.

2. On what does the boiling-point of water depend ?

3. Will the boiling-point be higher or lower at the top of a mountain ?   Why ?

4. Will it be higher or lower at the bottom of a mine ? Why ?

5. Describe an experiment showing the influence of a reduction of pressure upon the boiling-point.

6. Does water expand or contract in passing from the solid to the liquid state? Illustrate your reply by an experiment.

7. Name a substance that behaves in an opposite way from water in this respect.

8. Do substances expand or contract in passing from the liquid state into that of gas?

9. What space will be occupied by the steam from a cubic inch of boiling water?

IX. Other Effects of Heat—Freezing mixtures.—1. Give an instance of heat promoting chemical action.

2. Does chemical action generally produce heat?

3. Give an instance where the mixing of two things is accompanied with a lowering of temperature, and explain the result.

4. Why is a liquid that evaporates rapidly intensely cold?

5. Describe an experiment showing that water can be frozen by its own evaporation.

X. Distribution of Heat.—1. Has heat always a tendency to distribute itself?

2. In how many different ways will it do this?

3. Give an instance of *conduction;* of *convection;* of *radiation.*

XI. Conduction and Convection of Heat.—1. Describe an experiment showing that metal conducts heat better than glass.

2. Are wool and feathers good conductors or bad?

3. When do such bodies keep *in* heat?

4. When do such bodies keep *out* heat?

5. Describe an experiment showing that copper is a better conductor than iron.

6. What is the characteristic difference between conduction and convection?

7. Sketch the direction of the currents in a vessel of water heated beneath.

8. Explain the effect of convection in retarding the freezing of a lake.

9. Give an instance of the convection of air.

10. Explain the *trade* WINDS.

LIGHT FROM HEATED BODIES, p. 106.

I. Radiant Light and Heat—its Velocity.—1. By what process does the heat of the sun reach the earth?

2. Does a kettle containing hot water give out radiant heat?
PHYSICS.]

3. What sort of change takes place in the nature of the rays given out by a body as the process of heating it goes on ?

4. Who first found out the velocity with which light travels ?

5. Describe generally the manner in which the discovery was made.

6. At what rate does light travel ?

7. If the sun were suddenly extinguished, what time would elapse before we found it out ?

8. Does light consist of particles shot out ? If not, what is its nature ?

II. Reflexion of Light.—1. Illustrate the reflexion of light by an experiment.

2. Enunciate the law of reflexion in two statements.

3. Sketch a few letters of the alphabet and their images as given by a plane mirror.

4. What sort of image of external things have you in the bright bulb of a thermometer ?

5. Describe an experiment with two concave mirrors.

III. Bending or Refraction of Light.—1. Illustrate the bending of light by an experiment.

2. Sketch the direction of a ray of light before, during, and after its passage through a flat plate of glass.

3. Sketch the same when the glass is shaped like a wedge or prism.

4. Is the light bent *towards* or *from* the thickest part of the wedge ?

IV. Lenses—Images given by them.—1. Sketch a lens as it appears from above lying on the table.

2. Sketch a lens as it appears if viewed endwise.

3. Show the analogy between a lens and a prism.

4. Show by a sketch how a lens will bend a bundle of parallel rays from a distance falling upon it.

5. How may a lens be used as a burning glass ?

6. Show how a lens is used by a photographer.

V. Magnifying Glasses.—1. Show how a single lens may be used to magnify a small thing.

2. Will a single lens suffice if the thing be far away ?

3. In this case what arrangement would you adopt ? What is this called ?

VI. Different kinds of Light are differently Bent.—1. Suppose some blue, red, and green light fell together on a prism, would they emerge together ?

PHYSICS.]

2. If not, in what order would they be bent?

3. Of what colours mixed together is white light composed?

4. Give a sketch showing how we can prove this by means of the prism.

5. Who first discovered the compound nature of white light?

6. What is a spectrum? Illustrate your reply by reference to an experiment.

VII. Nature of Heat (second notice).

—1. If a blacksmith strikes a piece of lead with a heavy hammer what becomes of the energy of the blow?

2. What becomes of the energy spent in rubbing a button on a piece of wood?

3. Illustrate the conversion of the energy of a blow into heat by an experiment with a vesta.

4. Why do sparks fly out from the break-wheel of a railway train which is slackening its speed?

5. Give an instance where heat is changed back into visible energy.

ELECTRIFIED BODIES, p. 125.

I. Conductors and Non-Conductors.

—1. What was the first electrical fact known?

2. What discovery did Dr. Gilbert make?

3. Show by experiment that electricity cannot spread itself over glass.

4. Show by experiment that electricity can spread itself over metal.

5. What names are given to glass and metal in consequence of these properties?

6. Give a list of good and one of bad conductors.

II. Two kinds of Electricity.

—1. Describe an experiment showing that there are two kinds of electricity.

2. How do bodies behave to one another when charged with the same electricity? How when charged with opposite electricities?

3. Mention an experiment by which we separate the two kinds of electricity from each other.

4. When we rub a stick of glass with silk, with what kinds of electricity are these two substances electrified?

5. When we rub sealing-wax with flannel, with what kinds of electricity are these two substances electrified?

III. Action of Excited or Unexcited Bodies—Experiments.

—1. Explain by reference to an experiment what is meant by *electric induction*.

2. Describe and explain the electric spark.

3. Sketch the gold-leaf electroscope, and explain its action.

4. How will an electroscope charged with positive electricity be affected by an excited glass rod brought near its knob?

5. How by a stick of excited sealing-wax?

6. If you approach a reservoir of electricity with an insulated brass ball you get a small spark, but if the brass ball be connected with the earth you get a long spark. Why is this?

7. If a point be attached to the ball of last question you get no spark. Why is this?

8. What discovery was made by Franklin?

IV. Electrical Machine—Leyden Jar.—1. Roughly sketch the electrical machine, and describe its mode of action.

2. Sketch the Leyden jar, and describe its mode of action.

3. Sketch the discharging rod, and describe its use.

V. Energetic Nature of Electrified Bodies.—1. Give a proof that electricity is something which has energy in it.

2. In a flash of lightning is it electricity which you see? If not, what is it?

3. Why is an electric machine hard to turn?

VI. Electric Currents.—1. Sketch the battery of Volta, and describe its action.

2. What is meant by the pole-wires of a battery?

3. In what direction does the current of positive electricity pass through a completed circuit?

4. Sketch a Grove's battery, and describe its action.

VII. Properties of the Current.—1. How would you heat a platinum wire by an electric current?

2. How would you decompose water by such a current?

3. If water is decomposed, at what pole will the oxygen appear?—at what pole the hydrogen?

4. How can the electric current enable iron to attract iron?

5. Does soft iron retain this property after the current has ceased?

6. What is meant by a *magnet?*

7. How will a magnet place itself with reference to a current?

8. Explain how an electric telegraph becomes possible.

PHYSICS.]

DESCRIPTION OF APPARATUS.

No. of Experiment.		Price. £ s. d.
1, 2.—Tin pan, with peas		0 1 0
3.—Iron plate with four strings		0 1 6
4.—Balance to carry 2 lbs. in each scale ; beam two feet long		1 12 0
Piece of metal weighing 200 grains		0 1 0
Set of weights, 600 grains to ½ grain		0 10 6
5.—2 lbs. mercury in a bottle		0 10 0
Two pieces of glass two inches square		0 0 4
6.—Apparatus unnecessary.		
9, 10.—Beam of wood		0 1 9
Two 4-lb. weights		0 3 0
15.—Plumbline		0 1 0
Stoneware dish for mercury		0 0 6
16.—Tube for showing level of water		0 2 6
17.—Metal cylinder with two tubes and stoppers		0 6 0
Tube with moveable bottom and cord		0 3 0
Water-jar for tube		0 1 0
Indigo solution		0 0 5
18, 19.—Substance weighing 1,000 grains, same specific gravity as water		0 2 6
20.—Hollow brass cylinder		0 2 6
Bucket to contain it		0 2 6
Apparatus for attaching the bucket to balance		0 1 6
21.—See Experiment 18.		
22.—Block to illustrate flotation		0 0 3
24.—Apparatus unnecessary.		

No. of Experiment.		£	s.	d.
25.—Tate's air-pump		3	13	6
Bell-jar receiver		0	2	6
Two india-rubber balls		0	0	3
26.—Jar with neck and flange		0	2	6
Two pieces of india-rubber for it		0	1	0
27, 28, 29.—Box with strings		0	0	6
30.—Magdeburg hemispheres		0	5	6
Brass cock for ditto		0	3	0
31.—Tube for barometer		0	1	0
Glass mortar for cistern		0	1	0
Funnel for filling barometer		0	0	2
33.—Vibrating wire on support		0	1	0
37.—Model thermometer		0	5	0
Centigrade thermometer		0	4	0
38.—Bladder two-thirds filled with air		0	0	6
39.—Further apparatus unnecessary.				
40.—Use tin pan of Experiment 1.				
41.—Use flask of Experiment 42.				
42.—Flask for boiling water, and cork in duplicate		0	3	0
Triangle and wire gauze to support flask		0	1	5
43, 44.—No special apparatus necessary.				
45.—Pan to hold sulphuric acid *in vacuo*, and shallow vessel to hold water		0	3	8
46.—No apparatus necessary.				
47.—Use flask of Experiment 42.				
48.—Wires to show unequal power of iron and copper to conduct heat		0	1	0
50.—Use tin pan of Experiment 1.				
51.—Apparatus to show image of candle		0	10	6
52.—Apparatus unnecessary.				
54.—Electric pendulum		0	2	0
Several pieces of elder-pith		0	0	6
55.—Electroscope		0	12	0

No. of Experiment.		Price.		
		£	s.	d.
	Electrical machine, 16-inch plate	4	4	0
	Box of amalgam	0	1	0
56.—	Rod, half brass, half glass	0	2	6
	Rod of glass covered with red wax . . .	0	2	6
	Piece of silk	0	0	6
	Piece of flannel	0	0	6
57.—	No additional apparatus.			
58, 59.—	Brass ball, with point, on insulated stand .	0	3	6
60.—	No apparatus required.			
61.—	Leyden jar, pint size	0	4	0
	Discharger	0	3	0
62.—	Grove's battery, 4 cells in frame	1	18	0
	Yard of fine platinum wire	0	0	6
63.—	Voltameter	0	10	6
64.—	Electro-magnet	0	6	0
65.—	Knitting-needle and thread	0	0	2
66.—	Apparatus for Oersted's experiment . . .	0	5	6
67.—	Thirty feet of covered wire	0	1	3
		£19	3	8

As a wish has been expressed to have a cheaper if less complete set of apparatus, the following is offered as an alternative list.

DESCRIPTION OF APPARATUS.

No. of Experiment.		Price.		
		£	s.	d.
1, 2.—	Tin pan, with peas (supplied by experimenter)	0	0	0
3.—	Iron plate with four strings	0	1	6
4.—	Balance to carry 2 lbs. in each scale ; beam two feet long	1	12	0
	Piece of metal weighing 200 grains . . .	0	1	0
	Set of weights, 600 grains to ½ grain . . .	0	10	6
5.—	2 lbs. mercury in a bottle	0	10	0
	Two pieces of glass two inches square . .	0	0	4
	Carried forward . .	2	15	4

No. of Experiment.		Price. £	s.	d.
	Brought over . .	2	15	4
6.—Apparatus unnecessary.				
9, 10.—Beam of wood (supplied by experimenter).		0	0	0
	Two 4-lb. weights (ditto)	0	0	0
15.—Plumbline		0	1	0
	Stoneware dish for mercury	0	0	6
16.—Tube for showing level of water		0	2	6
17.—Metal cylinder with two tubes and stoppers				
	(not supplied)	0	0	0
	Tube with moveable bottom and cord . .	0	3	0
	Water-jar for tube	0	1	0
	Indigo solution	0	0	5
18, 19.—Substance weighing 1,000 grains, same specific gravity as water		0	2	6
20.—Hollow brass cylinder		0	2	6
	Bucket to contain it	0	2	6
	Apparatus for attaching the bucket to balance	0	1	6
21.—See Experiment 18.				
22.—Block to illustrate flotation		0	0	3
24.—Apparatus unnecessary.				
25.—Air-pump		2	2	0
	Bell-jar receiver	0	2	6
	Two india-rubber balls	0	0	3
26.—Jar with neck and flange		0	2	6
	Two pieces of india-rubber for it	0	1	0
27, 28, 29.—Box with strings		0	0	6
30.—Magdeburg hemispheres		0	5	6
	Brass cock for ditto	0	3	0
31.—Tube for barometer		0	1	0
	Glass mortar for cistern	0	1	0
	Funnel for filling barometer	0	0	2
33.—Vibrating wire on support		0	1	0
37.—Model thermometer		0	5	0
	Centigrade thermometer	0	4	0
38.—Bladder two-thirds filled with air		0	0	6
39.—Further apparatus unnecessary.				
40.—Use tin pan of Experiment 1.				
41.—Use flask of Experiment 42.				
42.—Flask for boiling water, and cork in duplicate		0	3	0
	Carried forward . .	7	5	11

No. of Experiment.		Price. £	s.	d.
	Brought over . .	7	5	11
	Triangle and wire gauze to support flask .	0	1	5
43, 44.—No special apparatus necessary.				
45.—This experiment cannot be shown with the air-pump of this list		0	0	0
46.—No apparatus necessary.				
47.—Use flask of Experiment 42.				
48.—Wires to show unequal power of iron and copper to conduct heat		0	1	0
50.—Use tin pan of Experiment 1.				
51.—Apparatus to show image of candle . . .		0	10	6
52.—Apparatus unnecessary.				
54.—Electric pendulum		0	2	0
Several pieces of elder-pith		0	0	6
55.—Electroscope		0	7	6
Electrical machine		1	10	0
Box of amalgam		0	1	0
56.—Rod, half brass, half glass		0	2	6
Rod of glass covered with red wax . . .		0	2	6
Piece of silk		0	0	6
Piece of flannel		0	0	6
57.—No additional apparatus.				
58, 59.—Brass ball, with point, on insulated stand .		0	3	6
60.—No apparatus required.				
61.—Leyden jar, pint size		0	4	0
Discharger		0	3	0
62.—Battery		0	12	6
Yard of fine platinum wire		0	0	6
63.—This experiment cannot be shown with the battery of this list		0	0	0
64.—Electro-magnet		0	6	0
65.—Knitting-needle and thread		0	0	2
66.—Apparatus for Oersted's experiment . . .		0	5	6
67.—Thirty feet of covered wire		0	1	3
		£12	2	3
	Price for the whole of the apparatus . .	£11	0	0

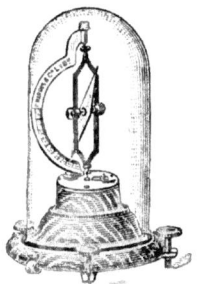
I

SCIENCE PRIMERS.

UNDER THE JOINT EDITORSHIP OF
Profs. HUXLEY, ROSCOE, AND BALFOUR STEWART.
18mo.   Illustrated.   1s. each.

INTRODUCTORY PRIMER. By
Professor HUXLEY, F.R.S.

CHEMISTRY. By Sir HENRY E.
ROSCOE, F.R.S.   With Questions.

PHYSICS. By BALFOUR STEWART,
F.R.S.   With Questions.

PHYSICAL GEOGRAPHY. By
ARCHIBALD GEIKIE, F.R.S.   With Questions.

GEOLOGY. By ARCHIBALD GEIKIE,
F.R.S.

PHYSIOLOGY. By Professor M.
FOSTER, M.D., F.R.S.

ASTRONOMY. By J. N. LOCKYER,
F.R.S.

BOTANY. By Sir J. D. HOOKER,
K.C.S.I., F.R.S.

LOGIC. By W. STANLEY JEVONS,
F.R.S.

POLITICAL ECONOMY. By W.
STANLEY JEVONS, F.R.S.

Others to follow.

MACMILLAN AND CO., LONDON.

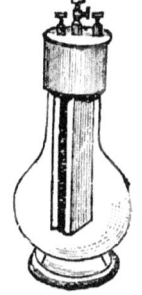
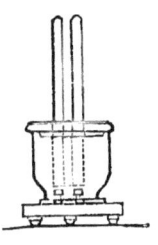

HISTORY AND LITERATURE PRIMERS.

18mo. 1s. each.

Edited by JOHN RICHARD GREEN.

ENGLISH GRAMMAR. By R. Morris, LL.D.
ENGLISH GRAMMAR EXERCISES. By R. Morris, LL.D., and H. C. Bowen, M.A.
EXERCISES ON MORRIS'S PRIMER OF ENGLISH GRAMMAR. By J. Wetherell, M.A.
ENGLISH COMPOSITION. By Prof. Nichol.
EXERCISES IN ENGLISH COMPOSI- TION. By Prof. Nichol. [*In the press.*
ENGLISH LITERATURE. By Stopford Brooke, M.A.
SHAKSPERE. By Prof. Dowden.
CHILDREN'S TREASURY OF LYRICAL POETRY. By F. T. Palgrave. In Two Parts, each 1s.
GREEK LITERATURE. By Prof. Jebb, Litt.D.
HOMER. By the Right Hon. W. E. Gladstone.
PHILOLOGY. By J. Peile, M.A.
GEOGRAPHY. By Sir George Grove, D.C.L. Maps.
CLASSICAL GEOGRAPHY. By H. F. Tozer, M.A.
GREEK ANTIQUITIES. By J. P. Mahaffy, M.A.
ROMAN ANTIQUITIES. By Prof. Wilkins, Litt.D., LL.D.
ROMAN LITERATURE. By A. S. Wilkins, Litt.D., LL.D.
EUROPE. By E. A. FREEMAN, D.C.L.
GREECE. By C. A. Fyffe, M.A. With Maps.
ROME. By M. Creighton, M.A. With Maps.
FRANCE. By C. M. Yonge. With Maps.

Others to follow.

MACMILLAN AND CO., LONDON.

4